UNDERSTANDING BACTERIOLOGY

UNDERSTANDING BACTERIOLOGY

By

Dr. Ashok Kumar

Dept. of Zoology
Bundelkhand University
Campus Department
Jhansi

D P H

DISCOVERY PUBLISHING HOUSE PVT. LTD.
NEW DELHI-110 002

First Published-2010

ISBN 978-81-8356-530-1

Published by:

DISCOVERY PUBLISHING HOUSE PVT. LTD.
4831/24, Ansari Road, Prahlad Street
Darya Ganj, New Delhi-110002 (India)
Phone: 23279245 • Fax: 91-11-23253475
E-mail: parul.wasan@gmail.com
info@discoverypublishinggroup.com
Website: www.discoverypublishinggroup.com

Printed at:

Sachin Printers
Delhi

Preface

The present title "Understanding Bacteriology" has been written for those students interested in careers in diverse fields of biological sciences. It provides a structured approach to learning by covering all the important topics in a uniform, systematic format. The book has been comprehensively designed incorporating recent advances in this fast moving field. It also provides accessible information on bacteriology in compact form for undergraduate students in biology and related life sciences. It is intelligible to the educated layman, though it deals with some complex ideas. It is an adequate text for all the requirements of students in this area. In addition, busy lecturers who require a quick reference compendium will find it useful, particularly for tutional planning. Simple, yet hopefully clear figures and tables are provided throughout the book.

The over-riding goal of this book, and indeed of the whole *Understanding series*, is to present the essential information concering bacteriology in a compact, readily accessible form which leads itself to student learning and revision. The convergence of various approaches has generated a rich panorama of detail, the significance of which we are still attempting to unraval. The present text has been written as an introduction to this rapidly growing field.

To make the work more comprehensive and informative, the author has consulted many authoritative books, research journals, abstracts, monographs etc., so there can be no claim to originality except in the manner of treatment.

The author expresses his thanks to his friends and colleagues whose continue inspirations have initiated him to bring out this book.

The author expresses his gratitude to Mr. Wasan and staff of M/s Discovery Publishing House Pvt. Ltd. for their whole hearted co-operation in the publication of this book.

In the mean time, the author will remain sincerely responsible for any shortcomings of the book and be grateful to the readers for their suggestions and constructive criticism for the continuous betterment of the book. He takes this opportunity to appeal to the readers to send their suggestions straightaway to his Publisher.

Author

CONTENTS

1

INTRODUCTION

As a group the *archaeobacteria* are quite diverse, both in morphology and physiology. They can stain either gram positive or gram negative and may be spherical, rod-shaped, spiral, lobed, plate-shaped, irregularly shaped, or pleomorphic. Some are single cells, whereas others are filaments or aggregates. They range in diameter from 0.1 to over 15 μm, and some filaments can grow up to 200 μm in length. Multiplication may be by binary fission, budding, fragmentation, or other mechanisms. Archaeobacteria are just as diverse physiologically. They can be aerobic, facultatively anaerobic, or strictly anaerobic. Nutritionally they range from chemolithoautotrophs to organotrophs. Some are mesophiles; other are hyperthermophiles that can grow above 100°C. Archaeobacteria usually prefer restricted or extreme aquatic and terrestrial habitats. They are often present in anaerobic, hypersaline, or high-temperature environments. Recently archaeobacteria have been discovered in cold environments. It appears that they constitute up to 34% of the prokaryotic biomass in coastal Antarctic surface waters. A few are symbionts in animal digestive systems.

Archaeobacterial Cell Walls

Although archaeobacteria can stain either gram positive or gram negative, their wall structure and chemistry differ form that of the eubacteria. There is considerable variety in archaeobacterial wall structure. Many gram-positive archaeobacteria have a wall with a single thick homogeneous layer like gram-positive eubacteria. Gram-negative archaeobacteria lack the outer membrane and complex peptidoglycan network or sacculus of gram-negative eubacteria. Instead they usually have a surface layer of protein of glycoprotein subunits.

The chemistry of archaeobacterial cell walls is also quite different form that of the eubacteria. None have the muramic acid and D-amino acids characteristic of eubacterial peptidoglycan. Not surprisingly, all archaeobacteria resist attack by lysozyme and β-lactam antibiotics such as penicillin. Gram-positive archaeobacteria can have a variety of complex polymers in their walls. *Methanobacterium* and some other methanogens have walls containing *pseudomurein*, a peptidoglycanlike polymer that has L-amino acids in its cross-links, N-acetyltalosaminuronic acid instead of N-acetylmuramic acid, and β(1→3) glycosidic bonds instead of β(1→4) glycosidic bonds. *Methanosarcina* and *Halococcus* contain complex polysaccharides similar to the chondroitin sulfate of animal connective tissue. Other heteropolysaccharides also found in grama-positive walls.

Gram-negative archaeobacteria have a layer of protein or glycoprotein outside their plasma membrane. The layer may be as thick as 20 to 40 nm. Sometimes there are two layers, a sheath surrounding an electron-dense layer. The chemical content of these walls varies considerably. Some methanogens (*Methanolobus*),

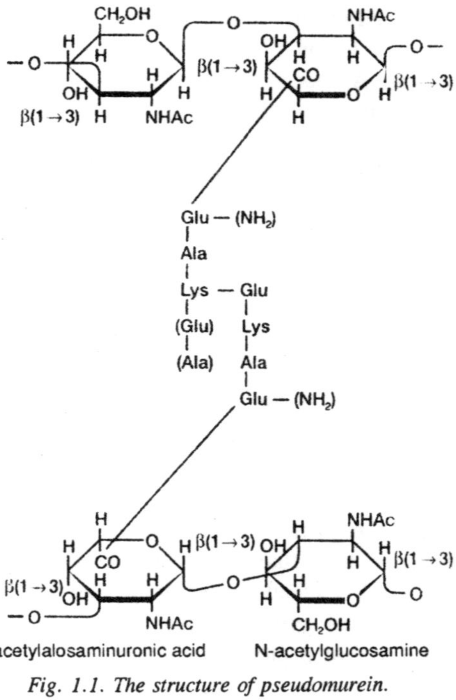

Fig. 1.1. The structure of pseudomurein.

Halobacterium, and several extreme thermophiles (*Sulfolobus*, *Thermoproteus*, and *Pyrodictium*) have glycoproteins in their walls. In contrast, other methanogens (*Methanococcus*, *Methanomicrobium*, and *Methanogenium*) and the extreme thermophile *Desulfurococcus* have protein walls.

Archaeobacterial Lipids and Membranes

Most distinctive feature of the archaeobacteria is the nature of their membrane lipids. They differ from both eubacteria and eukaryotes in having branched chain hydrocarbons attached to glycerol by ether links rather than fatty acids connected by ester links. Sometimes two glycerol groups are linked to form an extremely long tetraether. Usually the diether side chains are 20 carbons in size, and the tetraether chains are 40 carbons. The overall length of the tetraethers can be adjusted by cyclizing the chains to form pentacyclic rings, and biphytanyl chains may contain form one to four cyclopentyl rings. Polar lipids are also present in archaeobacterial membranes: phospholipids, sulfolipids, and glycolipids. From 7 to 30% of the membrane lipids are nonpolar lipids, which usually are derivatives of squalene. These lipids can be combined in various ways to yield membranes of different rigidity and thickness. For example, the C_{20} diethers can be used to make a regular bilayer membrane. A much more rigid monolayer membrane may be constructed of C_{40} tetraether lipids. Of course archaeobacterial membranes may contain a mix of diethers, tetraethers, and other lipids. As might be expected from their need for stability, the membranes of extreme thermophiles such as Thermoplasma and Sulfolobus are almost completely tetraether monolayers.

Genetics and Molecular Biology

Some features of archaeobacterial genetics are similar to those in eubacteria. Their chromosome is a single closed DNA circle. However, the genomes of some archaeobacteria are significantly smaller than the normal eubacterium. *E. coli* DNA has a size of about 2.5×10^9 daltons, whereas *Thermoplasma acidophilum* DNA is about 0.8×10^9 daltons and *Methanobacterium thermoautotrophicum* DNA is 1.1×10^9 daltons. The variation in G+C content is great, from about 21 to 68 mol% and is another sign of archaeobacterial diversity. Archaeobacteria have few plasmids.

Archaeobacterial mRNA appears similar of eubacteria rather than to eukaryotic mRNA. Polygenic mRNA has been discovered, and there is no evidence for mRNA splicing. Archaeobacterial promoters are similar to those in eubacteria.

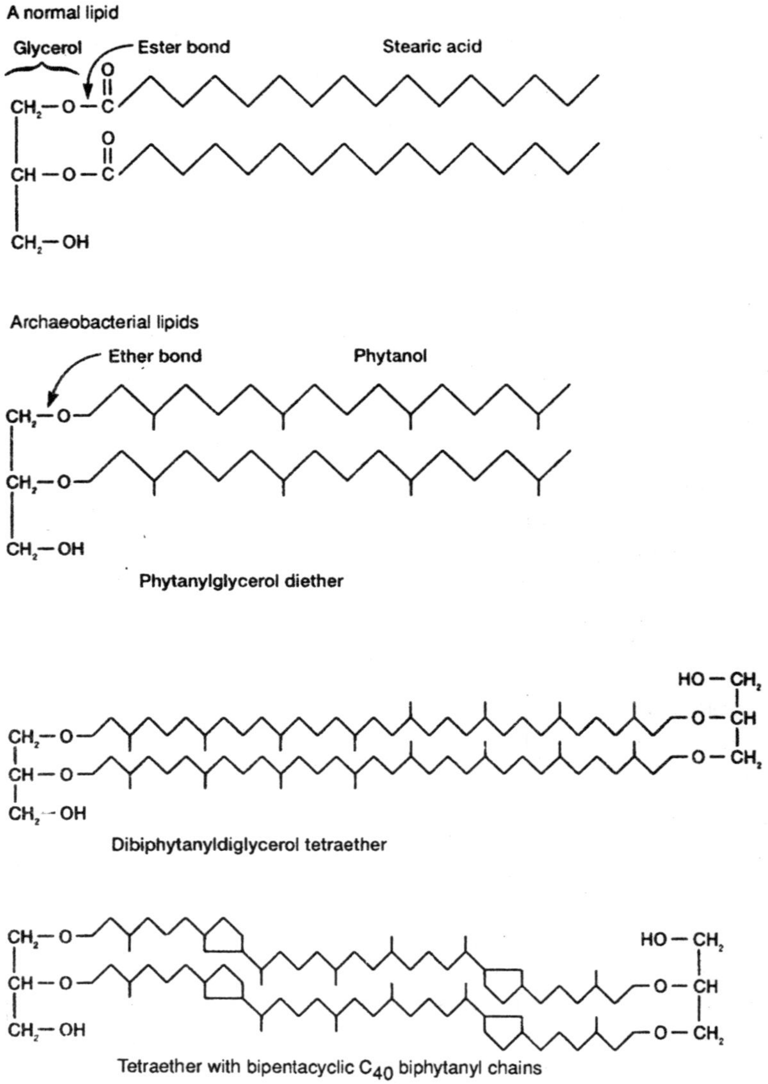

Fig. 1.2. Archaeobacterial membrane lipids.

Despite these and other similarities, there are also many differences between archaeobacteria and other organisms. Unlike both eubacteria and eukaryotes, the TψC arm of archaeobacterial tRNA lacks thymine and contains pseudouridine or 1-methylpseudouridine. The

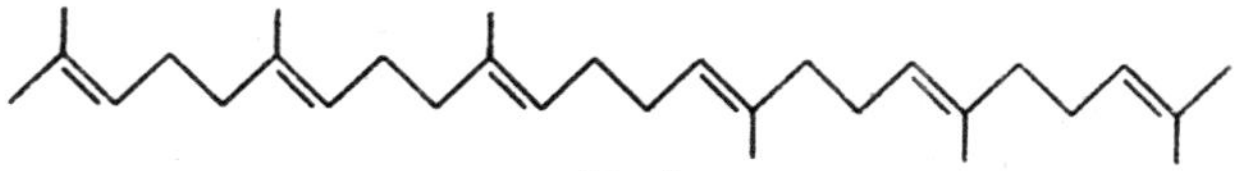

Squalene

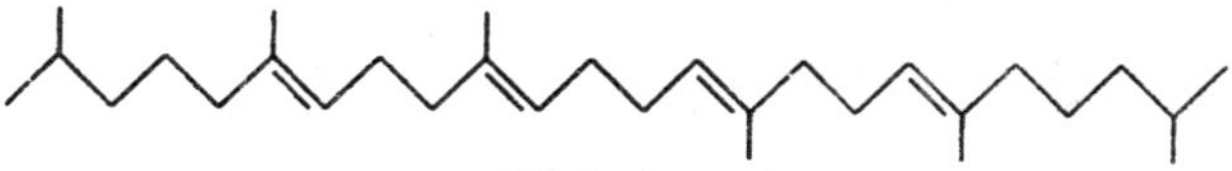

Tetrahydrosqualene

Fig. 1.3. Nonpolar lipids of archaeobacteria.

archaeobacterial initiator tRNA carries methionine like the eukaryotic initiator tRNA. Although archaeobacterial ribosomes are 70S like eubacterial ribosomes, electron microscopic studies show that their shape is quite variable and sometimes differs from that of both eubacterial and eukaryotic ribosomes. They do resemble eukaryotic ribosomes in their sensitivity to anisomycin and insensitivity to chloramphenicol and kanamycin. Furthermore, their elongation factor 2 reacts with diphtheria toxin like the eukaryotic EF-2. Finally, archaeobacterial DNA-dependent RNA polymerases resemble the eukaryotic enzymes, not the eubacterial RNA polymerase. They are large, complex enzymes and are insensitive to the drugs rifampin and streptolydigin. These and other differences distinguish the archaeobacteria from both eubacteria and eukaryotes.

Metabolism

Not surprisingly in view of the variety of their life-styles, archaeobacterial metabolism varies greatly between the members of different groups. Some archaeobacteria are organotrophs; others are autotrophic. A few even carry out an unusual form of photosynthesis.

Archaeobacterial carbohydrate metabolism is best understood. The enzyme 6-phosphofructokinase has not been found in archaeobacteria, and they do not appear to degrade glucose by way of the Embden-Meyerhof pathway. Extreme halophiles and thermophiles catabolize glucose using a modified form of the Enter-Doudoroff pathway in which the initial intermediates are not phosphorylated. The halophiles have slightly different modifications of the pathway than do the extreme thermophiles but still produce pyruvate and NADH or NADPH. Methanogens are autotrophs and do not catabolize glucose to any significant extent. In contrast with glucose degradation, gluconeogenesis proceeds by a reversal of the Embden-Meyerhof pathway in halophiles and methanogens. All archaeobacteria that have been studied can oxidize pyruvate to acetyl-CoA. They lack the pyruvate dehydrogenase complex present in eukaryotes and respiratory eubacteria and use the enzyme pyruvate oxidoreductase for this purpose. Halophiles and the extreme thermophile *Thermoplasma* do seem to have functional tricarboxylic acid cycle. No methanogen has yet been found with a complete tricarboxylic and cycle. Evidence for functional cytochrome chains has been obtained in halophiles and thermophiles.

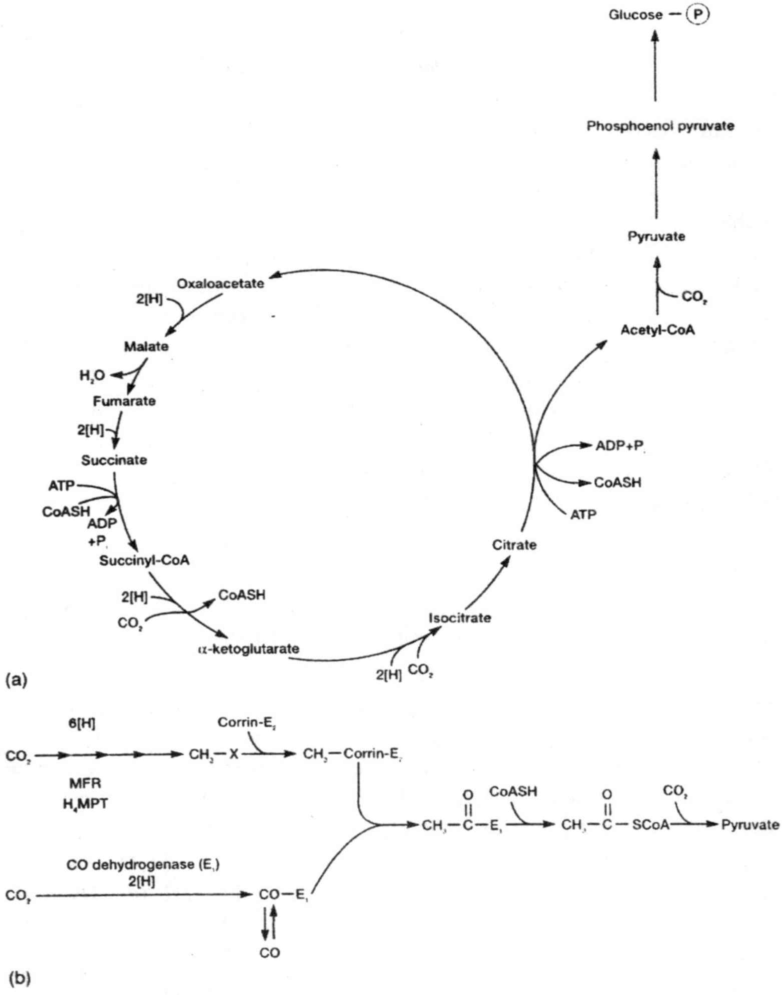

Fig. 1.4. Mechanisms of autotrophic CO_2 fixation.

Very little is known in detail about biosynthetic pathways in the archaeobacteria. Preliminary data suggest that the synthetic pathways for amino acids, purines, and pyrimidines are similar to those in other organisms. Some methanogens can fix atmospheric dinitrogen. Not only do many archaeobacteria use a reversal of the Embden-Meyerhof pathway to synthesize glucose, but at least some methanogens and extreme thermophiles employ glycogen as their major reserve material.

Autotrophy is widespread among the methanogens and extreme thermophiles, and CO_2 fixation occurs in more than one way. Thermoproteus and possibly Sulfolobus incorporate CO_2 by the reductive tricarboxylic acid cycle. This pathway is also present in the green sulfur bacteria. Methanogenic bacteria and probably most extreme thermophiles incorporate CO_2 by the reductive acetyl-CoA pathway. A similar pathway also is present in acetogenic bacteria and autotrophic sulfate-reducing bacteria.

Archaeobacterial Taxonomy

The archaeobacteria are quite distinct from other living organisms. Within the group, however, there is great diversity. Section 25 of *Bergey's Manual* divides the archaeobacteria into five major groups: methanogenic archaeobacteria, archaeobacterial, cell wall-less archaeobacteria, and extremely thermophilic S^0-metabolizers.

METHANOGENIC ARCHAEOBACTERIA

Methanogens are strict anaerobes that obtain energy by converting CO_2, H_2, formate, methanol, acetate, and other compounds to either methane or methane and CO_2. They are autotrophic when growing on H_2 and CO_2. This is the largest group of archaeobacteria. There are at least three orders and 18 genera, which differ greatly in overall shape, 16S rRNA sequence, cell wall chemistry and structure, membrane lipids, and other features. For example, methanogens construct three different types of cell walls. Several genera have walls with pseudomurein as mentioned earlier. Other walls contain either proteins or heteropolysaccharides.

As might be inferred from the methanogens' ability to produce methane anaerobically, their metabolism is unusual. These bacteria contain several unique cofactors: tetrahydromethanopterin (H_4MPT), methanofuran (MFR), coenzyme M (2-mercaptoethanesulfonic acid), coenzyme F_{420}, and coenzyme F_{430}. The first three cofactors bear the C_1 unit when CO_2 is reduced to CH_4. F_{420} carries electrons and hydrogens, and F_{430} is a nickel tetrapyrrole serving as a cofactor for

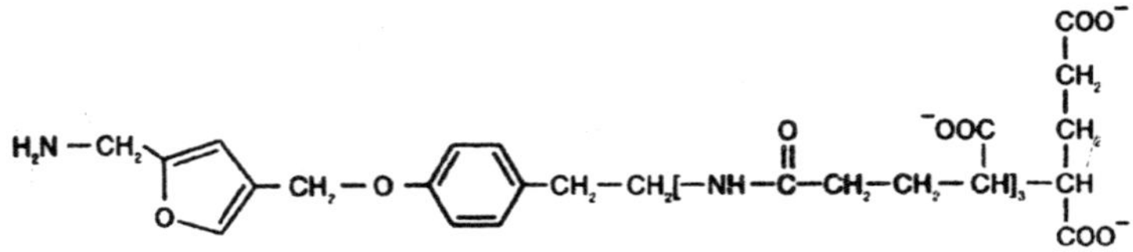

(a) Methanofuran (MFR)

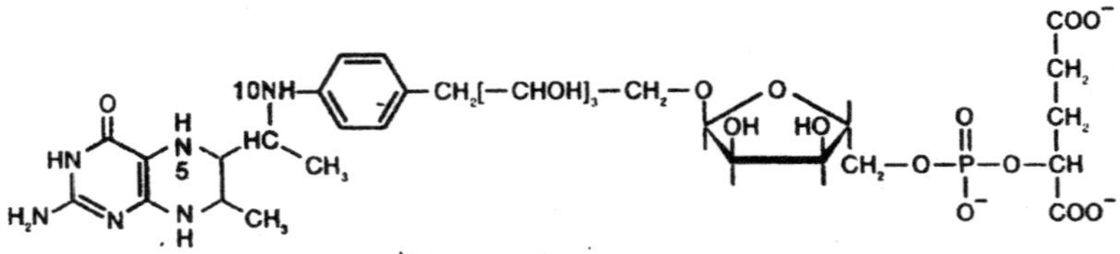

(b) Tetrahydromethanopterin (H_4MPT)

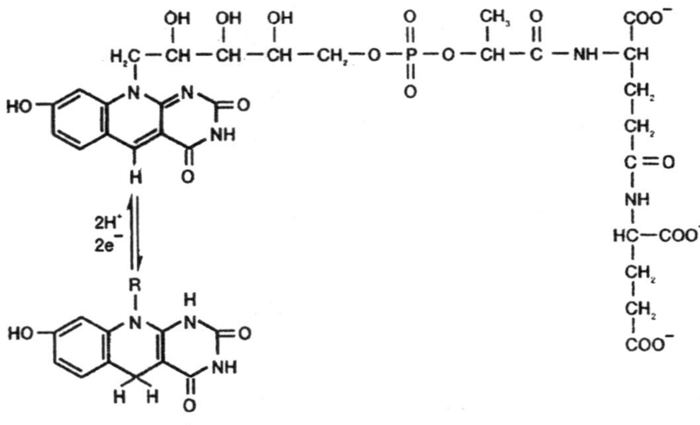

(c) Coenzyme F_{420}

Fig. 1.5. Methanogen coenzymes.

the enzyme methyl-CoM methylreductase. It appears that ATP synthesis is linked with methanogenesis by electron transport, proton pumping and a chemiosmotic mechanism. Some methanogens can live autotrophically by forming acetyl-CoA form two molecules of CO_2 and then converting the acetyl-CoA to pyruvate and other products.

Methanogens thrive in anaerobic environments rich in organic matter: the rumen and intestinal system of animals, freshwater and marine sediments, swamps and marshes, hot springs, anaerobic sludge

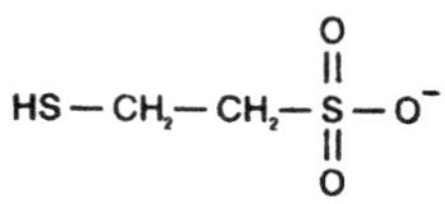

(a) Coenzyme M

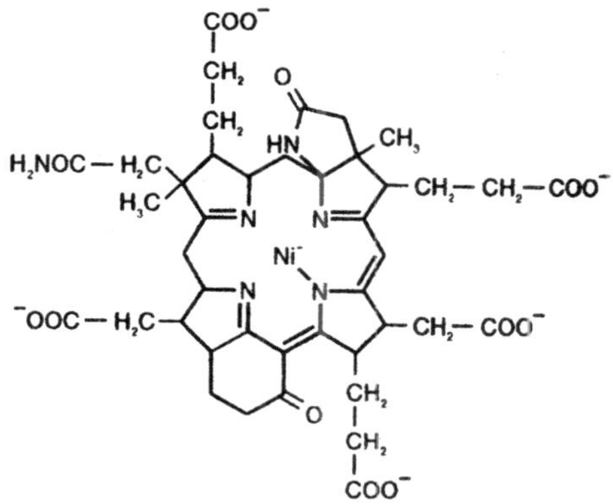

(b) Coenzyme F_{430}

Fig. 1.6. Methanogen coenzymes.

digesters, and even within anaerobic protozoa. Recently an extremely thermophilic rod-shaped methanogen has been isolated from a marine hydrothermal vent. *Methanopyrus kandleri* has a temperature minimum at 84°C and an optimum of 98°C; it will grow up to 110°C (above the boiling point of water). Methanogens often are of ecological significance. The rate of methane production can be so great that bubbles of methane will sometimes rise to the surface of a lake or pond. Rumen methanogens are so active that a cow can belch 200 to 400 liters of methane a day.

Methanogenic bacteria are potentially of great practical importance since methane is a clean-burning fuel and an excellent energy source. For many years sewage treatment plants have been using the methane they produce as a source of energy for heat and electricity. An anaerobic digester will degrade particulate wastes like sewage sludge to H_2, CO_2, and acetate. CO_2-reducing methanogens form CH_4 from CO_2 and H_2, while aceticlastic methanogens cleave acetate to CO_2 and CH_4 (about 2/3 of the methane produced by an anaerobic digester comes from acetate). A kilogram of organic matter can yield up to 600 liters of methane. It is quite likely that future research will greatly increase the efficiency of methane production and make methanogenesis an important source of pollution-free energy.

Methanogenesis also can be an ecological problem. Methane absorbs infrared radiation and thus is a greenhouse gas. There is evidence that atmospheric methane concentrations have been rising over the last 200 years. Methane production may significantly promote future global warming. Recently it has been discovered that methanogens can oxidize

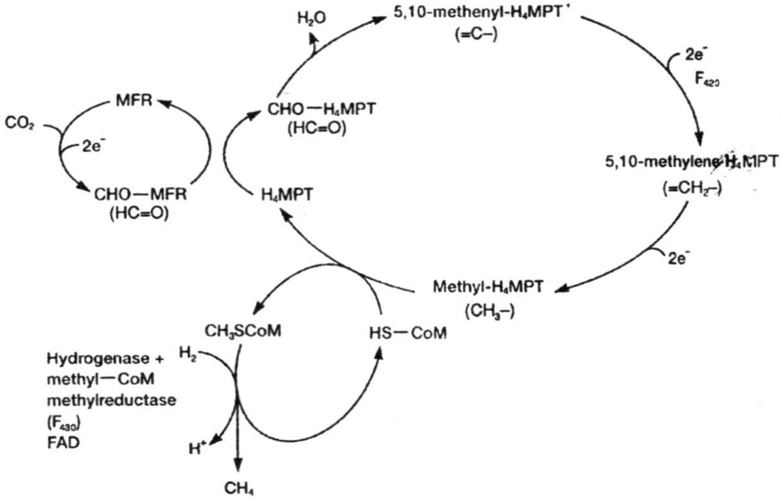

Fig. 1.7. Methane synthesis.

Fe^0 and use it to produce methane and energy. This means that methanogens growing around buried or submerged iron pipes and other objects may contribute significantly to iron corrosion.

Archaeobacterial Sulfate Reducers

This group contains gram-negative, irregular coccoid cells with walls consisting of glycoprotein sub units. At present, only the genus *Archaeoglobus* is known. It can extract electrons from a variety of electron donors (e.g., H_2, lactate, glucose) and reduce sulfate, sulfite, or thiosulfate to sulfide. Elemental sulfur is not used as an acceptor. *Archaeoglobus* is extremely thermophilic (the optimum is about 83°C) and can be isolated from marine hydrothermal vents. The organism is not only unusual in being able to reduce sulfate, unlike other archaeobacteria, but it also possesses the methanogen coenzymes F_{420} and methanopterin. It seems intermediate between the methanogens and the extremely thermophilic S^0-metabolizers.

Extremely Halophilic Archaeobacteria

The *extreme halophiles* are a third major group of archaeobacteria, currently with six genera in one family, the *Halobacteriaceae*. They are aerobic chemoheterotrophs with respiratory metabolism and require complex nutrients, usually proteins and amino acids, for growth. They are either nonmotile or motile by lophotrichous flagella. The most obvious distinguishing trait of this family is its absolute dependence on a high concentration of NaCl. These bacteria require at least 1.5M NaCl (about 8%, wt/vol), and usually have a growth optimum at abut 3 to 4 M NaCl (17 to 23%). They will grow at salt concentrations approaching saturation (about 36%). *Halobacterium's* cell wall is so dependent on the presence of NaCl that it disintegrates when the NaCl concentration drops to about 1.5 M. Thus halobacteria only grow in high-salinity habitats such as marine salterns and salts lakes like the Dead Sea between Israel and Jordan, and the Great Salt Lake in Utah. They also can grow in food products such as salted fish and cause spoilage. Halobacteria often have red-to-yellow pigmentation from carotenoids that are probably used as protection against strong sunlight. They can reach such high population levels that salt lakes, salterns, and salted fish actually turn red.

Probably the best-studied member of the family is *Halobacterium salinarium* (*H. halobium*). This prokaryote is unusual because it can trap light energy photosynthetically without the presence of chlorophyll. When exposed to low oxygen levels, some strains of *Halobacterium* synthesize a modified cell membrane called the *purple membrane*, which contains the protein *bacteriorhodopsin*. ATP is produced by a unique type of photosynthesis without the participation of bacterio-chlorophyll or chlorophyll, *Halobacterium* actually has four rhodopsin, each with a different function. As already mentioned, bacteriorhodopsin drives outward proton transport for purposes of ATP synthesis. Halorhodopsin uses light energy to transport chloride ions into the cell and maintain a 4 to 5 M intracellular KCl concentrations. Finally there are two rhodopsins that act as photoreceptors, one for red light and one for blue. They control flagellar activity to position the bacterium optimally in the water column. *Halobacterium* moves to a location of high light intensity, but one in which ultraviolet light is not sufficiently intense to be lethal.

Cell Wall-less Archaeobacteria

Thermoplasma grows in refuse piles of coal mines. These piles contain large amount of iron pyrite (FeS), which is oxidized to sulfuric acid by chemolithotrophic bacteria. As a result of the piles become very hot and acidic. This is an ideal habitat for *Thermoplasma* since it grows best at 55 to 59°C and pH 1 to 2. Although it lacks a cell wall, its plasma membrane is strengthened by large quantities of diglycerol tetraethers, lipopolysaccharides, and glycoproteins. The organism's DNA is stabilized by association with a special histonelike protein that condenses the DNA into particles resembling eukaryotic nucleosomes. At 59°C, *Thermoplasma* takes the form o: an irregular filament, whereas at lower temperature it is spherical. The cells may be flagellated and motile.

Extremely Thermophilic S^0-Metabolizers

The last group of archaeobacteria contains extremely thermophilic bacteria, many of which are acidophiles and sulfur dependent. The sulfur may be used either as an electron acceptor in anaerobic respiration or as an electron source by lithotrophs. Almost all are strict anaerobes. They grow in geothermally heated water or soils that contain elemental sulfur. These environments are scattered all over the world. Examples are the sulfur-rich hot springs in Yellowstone National Park, Wyoming, and the waters surrounding areas of submarine volcanic activity. Such habitats are sometimes called solfatara. These archaeobacteria can be very thermophilic and often are classified as hyperthermophiles. The most extreme example is *Pyrodictium*, a bacterium isolated from geothermally heated sea floors. *Pyrodictium* has a temperature minimum of 82°C, a growth optimum at 105°C, and a maximum at 110°C. Both organotrophic and lithotrophic growth occur in this group. Sulfur and H_2 are the most common electron sources for lithotrophs. There are three orders (*Thermococcales*, *Thermoproteales*, and *Sulfolobales*) and at least nine genera. Two of the better-studied genera are *Thermoproteus* and *Sulfolobus*.

Members of the genus *Sulfolobus* are gram-negative, aerobic, irregularly lobed spherical bacteria with a temperature optimum around 70 to 80°C and a pH optimum of 2 to 3. For this reason, they are classed as *thermoacidophiles*, so called because they grow best at acid pH values and high temperatures. There cell wall contains lipoprotein and carbohydrate but lacks peptidoglycan. They grow lithotrophically on sulfur granules in hot acid springs and soils while oxidizing the sulfur to sulfuric acid. Oxygen is the normal electron acceptor, but ferric iron may be used. Sugars and amino acids such as glutamate also serve as carbon and energy sources.

Thermoproteus is a long thin rod that can be bent or branched. The cell wall is composed of glycoprotein. *Thermoproteus* is a strict anaerobe and grows at temperatures from 70 to 97°C and pH values between 2.5 and 6.5. It is found in hot springs and other hot aquatic habitats rich in sulfur. It can grow organotrophically and oxidize glucose, amino acids, alcohols, and organic acids with elemental sulfur as the electron acceptor. That is, *Thermoproteus* can carry out anaerobic respiration. It will also grow chemolithotrophically using H_2 and S^0. Carbon monoxide or CO_2 can serve as the sole carbon source.

2

DIVERSITY OF BACTERIA

The art of biological classification is known as *taxonomy*. It has two functions: the first is to identify and describe as completely as possible the basic taxonomic units, or *species*; the second, to devise an appropriate way of arranging and cataloging these units.

SPECIES : THE UNITS OF CLASSIFICATION

The notion of a species is complex. Speaking broadly, a species consists of an assemblage of individuals (or, in microorganisms, of clonal populations) that share a high degree of phenotypic similarity, coupled with an appreciable dissimilarity form other assemblages of the same general kind. The recognition of species would not be possible if natural variation were continuous, so that an intergrading series spanned the gap between two assemblages of markedly different phenotype. However, it became evident early in the development of biology that, among most groups of plants and animals, reasonably sharp discontinuities do separate the members of a group into distinguishable assemblages. Hence, the notion of the species as the base of taxonomic operation proved workable.

Every assemblage of individuals shows some degree of internal phenotypic diversity, because genetic variation is always at work. Hence, it becomes a matter of scientific tact to decide what *degree* of phenotypic dissimilarity justifies the breaking up of an assemblage into two or more species; or, to put the matter another way, how much internal diversity is permissible is a species. Opinion on this question vary. Taxonomists themselves can be broadly divided into two groups: "lumpers," who set wide limits to a species, and "splitters," who differentiate species, on more slender grounds.

For plants and animals that reproduce sexually, a species can be defined in genetic and evolutionary term. As long as a sexually reproducing population is free to interbreed at random its total gene pool undergoes continuous redistribution, and new mutations, the source of phenotypic variation, are dispersed throughout the population. Such an interbreeding population may evolve in response to environmental changes, but it will evolve with reasonable uniformity. *Divergent evolution*, eventually leading to the emergence of new species, can occur only if a segment of the population becomes reproductively isolated in an environment that is different from that occupied by the rest of the population. Reproductive isolation is probably usually geographic in the first instance; a physical barrier of some sort (for example, a mountain range or a body of water) is interposed between two parts of the initially continuous, population. Within each of these subpopulations, a common gene pool is maintained by interbreeding, but through chance mutation and selection, the two subpopulation are now free to evolve along different lines. They will continue to diverge, as long as the geographical barrier persists. Eventually, the cumulative differences become so great that *physiological* isolation is superimposed on geographic isolation; members of the two populations are no longer capable of interbreeding if they are brought together. Hence, even if the two populations subsequently commingle once more, their gene pools remain permanently separated; a point of no return has been reached. These evolutionary considerations lead to dynamic definition of the species as a stage in evolution at which actually or potentially interbreeding arrays have become separated into two or more arrays physiologically incapable of interbreeding. This definition is in fact, an *explanation* of the origin of specific discontinuities in nature. At the same time, it provides an experimental criterion for the recognition of species differences: inability to interbreed.

Because most microorganisms are haploid, and reproduce predominantly by asexual means, the concept of the species that has emerged from work with plants and animals is evidently inapplicable to them. A microbial species cannot be considered an interbreeding population: the two offspring produced by the division of a bacterial cell are reproductively isolated from one another, and, in principle, they are free to evolve in a divergent manner. Genetic isolation is to some degree reduced by sexual or parasexual recombination in eukaryotic microorganisms and by the special mechanisms of recombination distinctive of bacteria. However, it is very difficult to assess the evolutionary effect of these recombinational processes, because the frequencies with which they occur in nature are unknown. In bacteria the problem is further complicated by plasmid transfer, which is relatively nonspecific, and permits exchanges of genetic material among bacteria of markedly different genetic constitution.

Since the dynamics of microbial evolution are so unlike the dynamics of evolution of plants and animals, there is not theoretical basis for the assumption that microbial evolution has led to phenotypic discontinuities that would justify the recognition of species. However, the experience of microbial taxonomists has shown that when many strains of a given microbial group are thoroughly analyzed, they can usually be divided into a series of discontinuous clusters: it is such *clusters of strains* that the microbial taxonomist recognizes empirically as species. Further insights into the dynamics of microbial evolution may eventually permit a format definition of the microbial species; if so, this will most likely be different from the species definition applicable to plants and animals.

In bacterial populations genetic change can occur so rapidly by mutation that it would be unwise to distinguish species on the basis of differences in a small number of characters, governed by single genes. Accordingly, the best working definition of a bacterial species is a group of strains that show a high degree of overall phenotypic similarity and that differ form related strain groups with respect to many independent characteristics.

Characterization of Species

Ideally, species should be characterized by complete descriptions of their phenotypes or—even better—of their genotypes. Taxonomic practice falls far short of these ideals; in most biological groups, even the phenotypes are only fragmentarily described, and genotypic characterizations are incomplete.

As a general rule, the phenotypic characters that can be most easily determined are structural or anatomical ones that can be directly observed. For this reason, biological classification is still based, at most levels, almost entirely on structural properties. Virtually the only exception is the classification of bacteria. The extreme structural simplicity of bacteria offers the taxonomist too small a range of characters upon which to base adequate characterizations. Hence, the bacterial taxonomist has always been forced to seek other kinds of characters—biochemical, physiological, ecological—with which to supplement structural data. The classification of bacteria is based, to a far greater extent than that of any other biological groups, on *functional* attributes. Most bacteria can be identified only by finding out what they can do, not simply how they look.

This confronts bacterial taxonomists with an additional problem. To find out what a bacterium can do, they have to perform experiments with it. The number of possible experiments that can be performed is extremely large, and although all will reveal facts, the facts so revealed will not necessarily be taxonomically significant ones, in the sense of contributing to a differentiation of the organism under study from related assemblages. Consequently, bacterial taxonomists can never be sure that they have performed the right experiments for taxonomic purposes; they may well have failed to perform certain experiments that would have shown them significant clustering in a collection of strains, and therefore erroneously conclude that they are dealing with a continuous series. There is no obvious way to get around this difficulty, except to make phenotypic characterizations as exhaustive as possible. However, an emerging alternative may soon resolve these dilemma; the molecular techniques for characterizing bacterial *genotypes* provide a possible objective basis for defining a bacterial species.

Naming of Species

According to a convention known as the *binomial system of nomenclature*, every biological species bears a latinized name that consists of two words. The first word indicates the taxonomic group of immediately higher order, or *genus* (plural, *genera*) to which the species belongs, and the second word identifies it as a particular species of that genus. The first letter of the genetic (but not of the specific) name is capitalized, and the whole phrase is italicized: *Escherichia* (genetic name) *coli* (specific name). In contexts in which no confusion is possible, the generic name is often abbreviated to its initial letter: *E. coli*.

A rigid and complex set of rules governs biological nomenclature; the rules are designed to keep nomenclature as stable as possible. The specific name given to a newly recognized species cannot be changed unless it can be shown that the organism has previously been described under another specific name, in which case the older name is used because it has priority. Unfortunately, the same stability does not govern the generic half of the name, since the arrangement of related species into general is an operation that can be carried out in different ways and that often changes in the course of time as new information becomes available. For example, *E. coli* has in the past been placed in the genus *Bacterium*, as *Bacterium coli* and in the genus *Bacillus*, as *Bacillus coli*. These three names are synonyms, since they all refer to one and the same species. This consequence of the binomial system can be very confusing, and taxonomic descriptions usually list all such synonyms in order to minimize the confusion. Binomial nomenclature is sued for all biological groups except viruses. The virologists are currently divided over the best way to designate members of this group; some wish to extend the binomial system to the viruses, whereas other would prefer another system, which gives in coded form information about the properties of the organism.

In a bacterial taxonomy, when a new species is named, a particular strain is designated as the type strain. Type strains are preserved in culture collections; if one is lost, a *neotype strain*, which resembles as closely as possible the description of the type strain, is chosen. The type strain is important for nomenclatural purposes, since the specific name is attached to it. If other strains, originally included in the same species, prove on subsequent study to deserve recognition as separate species, they must receive new names, the old specific name resting with the type strain and related strains.

In the taxonomic treatment of a biological group, the individual species are usually grouped in a series of categories of successively higher order: genus, family, order, class, and division (or phylum). Such an arrangement is known as a *hierarchical* one, because each category in the ascending series unites a progressively larger number or taxonomic units in terms of a progressively smaller number of shared properties. It should be noted that the genus has a position of special importance, since according to the rules of nomenclature a species cannot be named unless it is assigned to a genus. The allocation of a species to a taxonomic category higher than the genus does not carry any essential *nomenclature* information, relative to other organisms, in the system of arrangement adopted.

Problems of Taxonomic Arrangement

In dealing with a large number of different objects, some system of orderly arrangement is essential for purposes of data storage and retrieval. It does not matter what criteria for making the arrangement are adopted, provided that they are unambiguous and convenient. Books can be arranged in different ways: for example, by subject, by author, or by title. Different individuals tend to adopt different systems, depending on their particular needs and tastes. Such a system of classification, based on arbitrarily chosen criteria, is termed an *artificial* one.

The earliest systems of biological classification were largely artificial in design. However, as knowledge about the anatomy of plants and animals increased, it became evident that these organisms conform to a number of *major patterns* or *types*, each of which shares many common properties, including ones that are not necessarily obvious upon superficial examination. Examples of such types are the mammalian, avian, and reptilian types among vertebrate animals. The first system of biological classification that attempted to group organisms in terms of such typological resemblances and differences was developed in the middle of the eighteenth century by Linnaeus. The Linnaean arrangement was more useful than previous artificial arrangements, since the taxonomic position of an organism furnished a large body of information about its properties: to say that an animal belongs to the vertebrate class Mammalia immediately tells one that it possesses all those properties which distinguish mammals collectively from other vertebrates. Because Linnaean classification expressed the *biological nature* of the objects that it classified, it became known as a *natural system* of classification, in contrast to preceding artificial systems.

Phylogenetic Approach to Taxonomy

When the fact of biological evolution was recognized, another dimension was immediately added to the concept of a natural classification. For biologists of the eighteenth century, the typological groupings merely expressed *resemblances*; but for post-Darwinian biologists, they revealed *relationships*. In the nineteenth century the concept of a "natural" system accordingly changed: it became one that grouped organisms in terms of their *evolutionary affinities*. The taxonomic hierarchy became in a certain sense the reflection of a family tree, and taxonomy suddenly acquired a new goal: the restructuring of hierarchies to mirror evolutionary relationships. Such a taxonomic system is known as a *phylogenetic system*.

Numerical Taxonomy

An alternative approach is an empirical one: the attempt to base taxonomic arrangement upon *quantification* of the similarities and differences among organisms. This was first suggested by Michel Adanson, a contemporary of Linnaeus, and is known as *Adansonian* (or *numerical*) *taxonomy*. The underlying assumption is that, provided each phenotypic character is given explicit weighting, it should be possible to express numerically the taxonomic distances between organisms, in terms of the number of characters they share, relative to the total number of characters examined. The significance of the numerical relationships so determined is greatly influenced by the number of characters examined; these should be as numerous and as varied as possible, to obtain a representative sampling of phenotype.

Until recently, the Adansonian approach appeared impractical because of the magnitude of the numerical operations involved. This difficulty has been obviated by the advent of computers, which can be programmed to compare data for a large number of characters and organisms and to compute the degrees of similarity. For any pair of organisms, the calculation of similarity can be made in two slightly different ways. The *similarity coefficient* S_J does not take into account characters negative for both organisms, being based only on positive matches; the *matching coefficient* S_S includes both positive and negative matches in the calculation.

The data can then be transposed into a *dendrogram* as a basis for determining taxonomic arrangement in terms of numerical relationships. The two dotted vertical lines indicate similarity levels that might be considered appropriate for recognizing two different taxonomic ranks (e.g., genus and species).

Numerical taxonomy does not have the evolutionary connotations of phylogenetic taxonomy, but it provides an objective and stable basis for the construction of taxonomic groupings. Perhaps its greatest advantage is that it cannot be applied at all until a relatively large number of characters have been determined, so that its use encourages a thorough examination of phenotypes. The analyses are open to

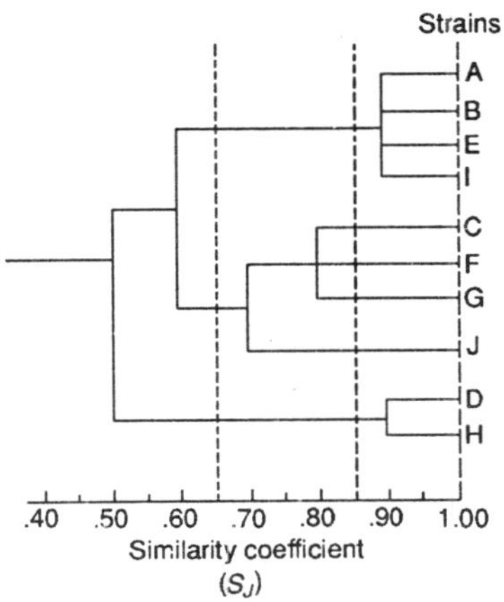

Fig. 2.1. A dendrogram showing similarily relationships among 10 bacterial strains.

continuous revision and refinement as more characters in a given group are determined.

New Approaches to Bacterial Taxonomy

The growth of molecular biology has opened up a number of new approaches to the characterization or organisms, which have had a profound impact on the taxonomy of bacteria. Of particular value are certain techniques that give insights into genotypic properties and thus complement the hitherto exclusively phenotypic characterization of these organisms. Several kinds of analysis performed upon isolated nucleic acids furnish information about genotype: the analysis of the base composition of DNA, the study of chemical hybridization between nucleic acids isolated from different organisms, and the sequencing of nucleic acids.

Base Composition of DNA: Its Determination and Significance

DNA contains for bases: adenine (A), thymine (T) guanine (G), and cytosine (C). For double-stranded DNA, the base-pairing rules require that A = T and G = C. However, there is no chemical restriction on the molar ratio (G + C) : (A + T). Early in the chemical study of DNA, analyses showed that this ratio in fact varies over a rather wide range in DNA preparations from different organisms, and subsequent work has revealed that the base composition of DNA is a character of profound taxonomic importance, particularly among microorganisms.

Although DNA base composition may be determined chemically, after hydrolysis of a DNA sample and separation of the free bases, it can be determined more easily by physical methods, and these are now the ones principally used. The "melting temperature" of DNA (i.e., the temperature at which it becomes denatured, by breakage of the hydrogen bonds that hold together the two strands) is directly related to G + C content, because hydrogen bonding between GC pairs is stronger than that between AT pairs (a consequence of the fact that GC pairs from three hydrogen bonds, while AT pairs form only two). Strand separation is accompanied by a marked increase in absorbance at 260 nm, the absorption maximum of DNA, and this can be easily measured in a spectrophotometer. When a DNA sample is gradually heated, the absorbance increases as the hydrogen bonds are broken and reaches a plateau at a temperature at which the DNA has all become single-stranded. The midpoint of this rise, the melting temperature (T_m) is a measure of the G + C content. The G + C content of DNA may also be determined by subjecting a DNA sample to

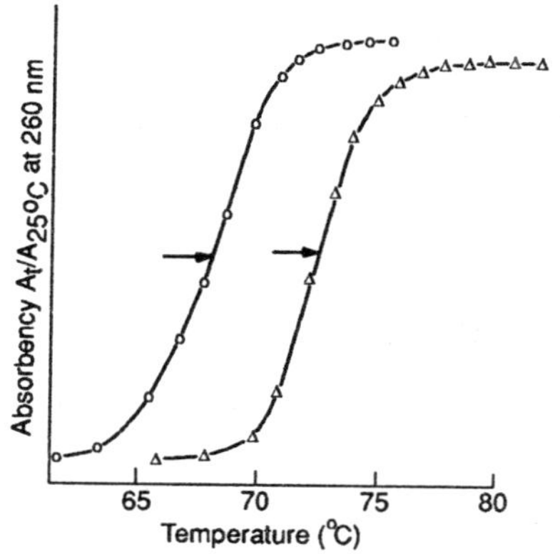

Fig. 2.2. Melting curves determined optically for two samples of bacterial DNA. Curve A–DNA of Lactobacillus acidophilus (T_m 67.7°C). Curve B–DNA of Leptospira sp. (T_m 72.1°C).

centrifugation in a CsCl gradient, and determining optically the position at which the DNA bands in the gradient, which affords a precise measure of its density. This method can be used because the density of DNA is also a function of the (G +C): (A + T) ratio.

Physical methods of analysis also provide an indication of the *molecular heterogeneity* of a DNA sample. If every molecule of DNA had the same G + C content, both the thermal transition in a melting

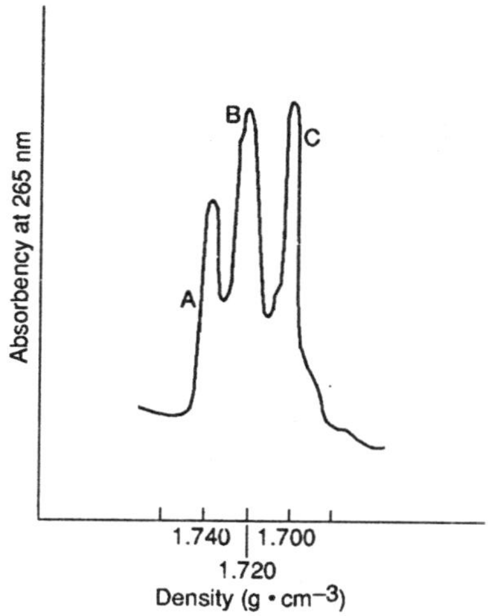

Fig. 2.3. The position in a CsCl density gradient assumed after centrifugation by three different DNAs of different G+C content. (A) DNA of a bacteriophage of Bacillus subtilis. (B) DNA of Thiobacillus novellus. (C) DNA of Leptospira sp

curve and the band position in a CsCl gradient would be extremely sharp. The steepness of the curve for thermal transition and the narrowness of the band in a gradient are therefore directly related to the homogeneity of G + C content in a population of DNA molecules. Even when DNA has been considerably fragmented by shearing an unavoidable consequence of normal handling of large DNA molecules like the bacterial chromosome), preparations from most organisms remain relatively homogeneous by these criteria, which indicates that the mean G + C content varies little in different parts of the genome. The only major exceptions are preparations from organisms that contain two genetic elements of different G + C content. Thus, in preparations from certain eukaryotic organisms, DNA of mitochondrial or chloroplast origin may differ appreciably in G +C content from the nuclear DNA, and there is sometimes a marked molecular heterogeneity in the DAN of a bacterium that harbors a plasmid. In such cases, the minor constituent may form a distinct *satellite band* in a CsCl gradient; this phenomenon provided one of the clues tat led to the discovery of DNA in mitochondria and chloroplasts.

Since no DNA preparation shows *absolute* molecular homogeneity, the G +C content is always a *mean* value and represents the peak in a normal distribution curve.

Taxonomic Implications of DNA Base Composition

In both plants and animals the ranges are relatively narrow and quite similar, centering about a value of 35 to 40 percent G + C. Among the protists the ranges are much wider. The widest range of all occurs among the prokaryotes, in which the range extends form about 25 to nearly 80 percent G + C. If, however, one examines the mean G + C content of many different strains that are assigned to a *single microbial species*, the values are closely similar or identical as shown by the data for several *Pseudomonas* species assembled. Each bacterial species, accordingly, has DNA with a characteristic mean G + C content; this can be considered one of its important specific characters. Furthermore, a substantial divergence between two organisms with respect to mean DNA base composition reflects a large number of individual differences between the specific base sequences of their respective DNAs. It is prima facie evidence for a major genetic divergences and hence for a wide evolutionary separation. The very broad span of values characteristic of the prokaryotes, accordingly, reveals the great evolutionary diversity of this particular biological group, and it also suggests its evolutionary antiquity.

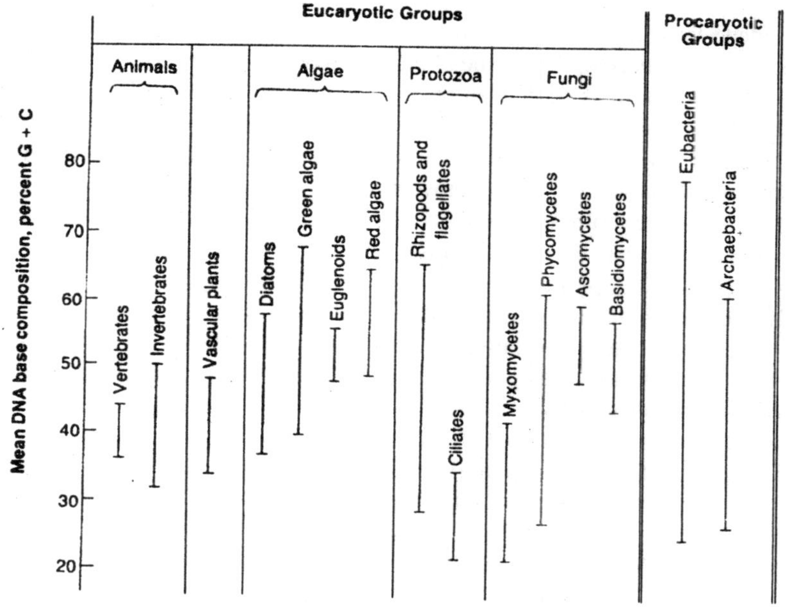

Fig. 2.4. The ranges of mean DNA base composition (percent G+C) characteristic of major biological groups.

However, two organisms with identical mean DNA base compositions may differ greatly in genetic constitution. This is evident from the very similar base ratio values for DNA from all plants and animals. Hence, major evolutionary divergence is not *necessarily* expressed by a divergence of mean base composition. When two organisms are closely similar in their DNA base composition, this fact can be construed as indicative of genetic and evolutionary relatedness only if the organisms also share a large number of phenotypic properties in common or are known to resemble one another in genetic constitution (e.g., different strains that belong to a single bacterial species). In such a case, near-identity of DNA base composition provides supporting evidence for their genetic and evolutionary relatedness.

Mean DNA base composition is a character of particular taxonomic value among bacteria, since the range for the group as a whole is so wide. Values have now been determined for a large number of representative strains and species belonging to every major subgroup of prokaryotes. Although the values for the constituent species in a genus differ somewhat the total range within a bacterial genus is in general fairly narrow (rarely greater than 10 to 15 percent), and can indeed be considered as an important character for the definition of a genus. Furthermore, the members of multigeneric clusters that had been recognized as similar on phenotypic grounds have been recognized as similar on phenotypic grounds have proved in some cases to be closely similar in DNA base composition. Thus, the fruiting myxobacteria and the euactinomycetes, two large bacterial multigeneric clusters, are both characterized by narrow spans of DNA base composition that lie near the high end of the G + C scale.

It has become evident, however, that some bacterial groups that had been closely associated by virtue of phenotypic resemblances are in fact unrelated genetically, as revealed by wide differences in DNA base composition. The gliding, unicellular nonphotosynthetic bacteria of the *Cytophaga* group, similar to fruiting myxobacteria with respect to vegetative cell structure, lie almost at the opposite end of the G + C scale; and the same situation occurs among the Gram-positive cocci of the *Staphylococcus* and *Micrococcus* groups.

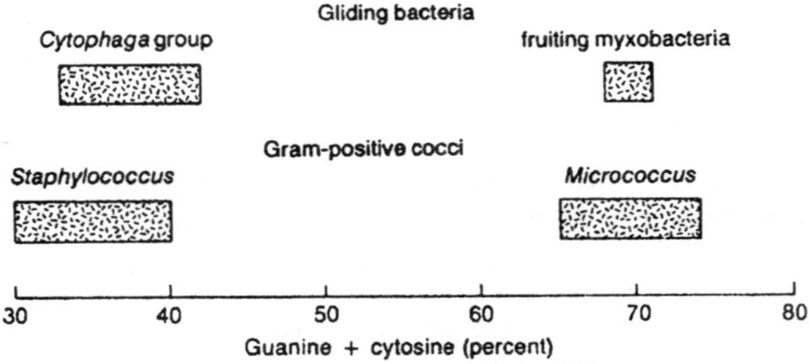

Fig. 2.5. Ranges of mean DNA base composition of certain representatives in two bacterial assemblages defined by phenotypic properties: gliding bacteria and Gram-positive cocci.

Nucleic Acid Hybridization

Upon rapid cooling of a solution of thermally denatured DNA, the single strands remain separated. However, if the solutionis held at a temperature from 10°C to 30° below the T_m value, specific reassociation ("annealing") of complementary strands to form double-stranded molecules occurs. There is always some random pairing, but since a randomly matched duplex contains many mismatched base pairs, its thermal stability is low and its strands separate very rapidly at temperatures near the T_m. In contrast, pairing of complementary strands forms duplexes that are quite stable because each base participates in interstrand hydrogen bonding. Thus at temperatures near the T_m, only duplexes between strands with a high degree of complementarily persist; the closer that the temperature of incubation is to the T_m, the more stringent is the requirement for precise base pairing.

Shortly after discovery of this phenomenon, it was shown that when DNA preparations from two related strains of bacteria are mixed and treated in this manner, hybrid DNA molecules are formed. One bacterial strain was grown in a medium containing D_2O, so that its DNA was "heavy" as a result of deuterium incorporation. After the two DNA samples had been mixed, denatured, and annealed hybrid molecules could be detected by centrifugation in a CsCl gradient, where they formed a band intermediate in position between those of the "light" and "heavy" duplex. In similar experiments, conducted with DNA preparations from two unrelated bacteria, no hybridization could be detected; upon annealing, duplexes were formed only by specific pairing between single strands originally derived from the same DNA.

The discovery of the reassociation of single-stranded DNA molecules from different biological sources to form hybrid duplexes laid the foundations of an entirely new approaches to the study of genetic relatedness in bacteria. *In vitro* experiments on DNA-DNA reassociation permit an assessment of the overall degree of genetic homology between two bacteria. Furthermore, since duplexes can also be formed between single-stranded DNA and complementary RNA strands, analogous DNA-RNA reassociations can be performed. If the RNA preparations consist of either tRNAs or rRNAs, such experiments permit an assessment of the genetic homology between two bacteria with respect to specific, relatively small segments of the chromosome: those that code the base sequences either of the transfer RNAs or of the ribosomal RNAs. As will be discussed later, DNA-rRNA similarities are of particular taxonomic interest.

Techniques and Interpretations of Reassociation Experiments

The density gradient method is too cumbersome for routine use, and a variety of simpler methods for measuring nucleic acid reassociation have been developed. All are based on the same general procedure: formation of duplexes between two denatured DNA samples, one of which is labeled with a radioisotope; separation of the duplexes from residual single-stranded nucleic acid; and measurement of their radioactivity. A point of reference is always required: it is provided by the DNA of a *reference strain*, which is prepared both labeled and unlabeled. The amount of reassociation between these two homologous

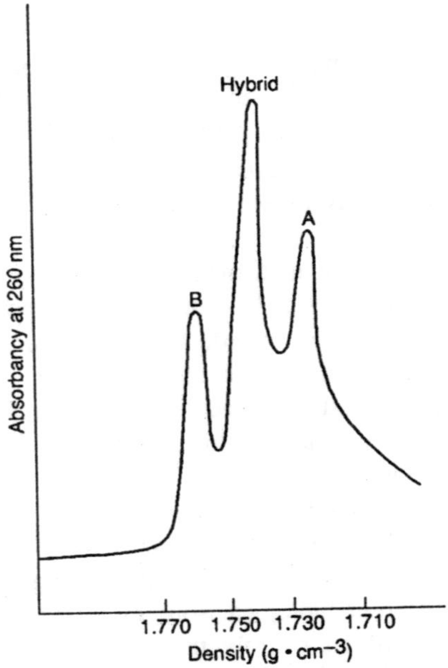

Fig. 2.6. An experimental demonstrating the formation of hybrid DNA molecules through reassociation of single-stranded (denatured) DNA molecules to form double helical DNA.

DNAs is determined and is assigned an arbitrary value of 100. The amount of reassociation between the reference DNA and DNAs from heterologous strains can then be measured and expressed as a percentage of the normalized value for the homologous DNA-DNA reassociation. Depending on the particular technique used, either the reference DNA or the heterologous DNA may be labeled. The same principles apply to DNA-RNA reassociation experiments.

In every type of reassociation experiment the temperature of reassociation, the ionic strength of the solvent, and the mean length of the DNA fragments must be standardized, since these factors all affect the rate of duplex formation. When they are held constant, reassociation is determined solely by the *DNA concentration* and the time of incubation. Provided that these two parameters are correctly chosen, nearly complete reassociation of complementary strands will occur.

The temperature at which reassociation is conducted is a factor of paramount importance. For obvious reasons, this temperature must always be below the T_m value of the reference DNA. At so-called *stringent* temperatures, 10° to 15°C below T_m, only complementary strands with a very high degree of base sequence homology can form stable duplexes. At somewhat lower temperatures, stable duplexes can also be formed between complementary strands in which base pairing is less perfect and hydrogen bonding therefore somewhat weaker.

If reassociation experiments are conducted at two different temperatures within the specific reassociation range (e.g., T_m - 15°C and T_m –30°C), the *degree of a base sequence homology* between a series of heterologous DNAs and the reference DNA can be roughly assessed. If homology is very high, reassociation is little affected by temperature; if base pairing is less perfect, reassociation will be markedly reduced at the higher temperature.

Carefully controlled DNA-DNA reassociation experiments can thus provide much semiquantitative information about the degree of genetic homology between related strains or species of bacteria. However, if evolutionary divergence has led to numerous differences of base sequence between the two genomes, specific DNA-DNA reassociation becomes too weak to be measured. The range of organisms among which genetic homology is detectable can be greatly extended by parallel studies on DNA-rRNA reassociation, because the relatively small portion of the bacterial genome that codes for ribosomal RNAs has a much more highly conserved base sequence than the bulk of the chromosomal DNA. As a result, it is frequently possible to detect by DNA-rRNA reassociation relatively high homology between the genomes of two bacteria which show no significant homology by DNA-DNA reassociation.

In any given bacterial group the value of nucleic acid reassociation studies is directly related to the number of strains and species that have been compared. Extensive comparative data are now available for several major bacterial groups. The insights that have been obtained into the genetic relationship among about 30 species of aerobic pseudomonads, currently classified in the genera *Pseudomonas* and *Xanthomonas*. Each large, shaded circle in this diagram embraces a single rRNA homology group. The species included in each group all show a much higher degree of relatedness with one another than with any species belonging to other rRNA homology groups. In fact, *intergroup* level of rRNA homology among aerobic pseudomonads are no greater than those with the enteric bacterium, *Escherichia coli*.

Within each rRNA homology group, the smaller, unshaded circles define the limits of inter specific genetic homology detectable by DNA-DNA hybridization. It can be seen that one of the rRNA homology group includes three isolated DNA homology groups (*Pseudomonas acidovorans* + *P. testosteroni*; *P. saccharophila*; *P. facilis* + *P. delafieldii*). In the four other rRNA homology groups, all the constituent species can be shown to be genetically interrelated by pairwise DNA-DNA hybridization experiments, even though some pairs of species within a given DNA homology group may be so distantly related that no DNA homology can be detected between them. This is true, for example, of the two species of fluorescent pseudomonads, *P. aeruginosa* and *P. syringae*; however, both show some degree of relatedness in terms of DNA-DNA reassociation with a third species, *P. fluorescens*. Another interesting point concerns the yellow-pigmented plant pathogenic pseudomonads, currently classified in a separate genus, *Xanthomonas*. Both DNA-rRNA and DNA-DNA hybridization studies show that they form a homology group with one otherwise genetically isolated *Pseudomonas* species, *P. maltophilia*.

Nucleic Acid Sequencing

Reassociation studies utilizing heterologous DNAs or RNAs can yield substantial information about the similarity of their base sequences; however, such studies cannot establish the actual order of bases in the molecule. Knowledge of the sequence of bases in nucleic acid would not only establish the similarity between two organisms, but would provide much additional information as well; for instance, the amino acid sequence of any protein could be read directly form the base sequence, and regions with regulatory function could be identified. Recently, methods have been developed for rapidly sequencing both DNA and RNA.

DNA Sequencing

Either of two different procedures can be used to sequence substantial lengths of DNA; the Sanger method or the *Maxam and Gilbert* method. Although they differ in chemical detail, both rest on the same basic principle: a series of DNA fragments are generated which have a common starting point, but variable termini ("nested" fragments); by determining their exact length and the terminal base, the sequence can be inferred with great accuracy. As an example, the Sanger method will be described.

In order to determine the sequence of bases in a segment of DNA, a large number of identical copies of it must be obtained. If the molecule that contains the sequence is relatively small (e.g., a viral genome or a plasmid), one can feasibly sequence the entire molecule. In such a case one need only separate the molecule from contaminating nucleic acid (e.g., fragments of the bacterial chromosome). If the sequence is part of a much larger molecule, such as the bacterial chromosome, the usual approach is to *clone* the desired segment, and sequence a restriction fragment purified from the cloned DNA. Since the Sanger method uses single-stranded DNA, the segment must either be digested with a nuclease to convert it to single-stranded DNA or, preferably, cloned in a single-stranded vector to begin with (such as the bacteriophage M13).

Sequencing of a purified single-stranded fragment is accomplished by generating a series of complementary nested fragments of DNA polymerase, a short primer sequence complementary to a region of the fragment (often a short restriction fragment itself), the four deoxy nucleoside triphosphates (one or more of which are radioactive), and a small amount of one *dideoxy* nucleoside triphosphate (e.g., dideoxythymidine, which lacks a 3'-hydroxyl group, terminates polymerization at that point. Thus, after a suitable incubation period, the mixture will contain radioactive DNA strands of variable length, all of which terminate in a thymidine residue. The number of classes of DNA molecules, differing from each other in length, depends on the number of thymidine residues, since for every thymidine residue in the new strand, there will be a family of radioactive DAN molecules that terminates at that point.

In practice four parallel incubations are performed, identical except that each contain a different dideoxy nucleoside triphosphate, and hence will terminate at a different base. The mixture is then denatured and electrophoresed to separate the newly synthesized strands by size. The position of the radioactive bands on the gel are visualized by appressing the gel tightly to a sheet of X-ray film, which is exposed by the localized decay of the radioisotope incorporated into the DNA. Upon development, the film displays a series of exposed bands, each of which corresponds to a size class of single-stranded DNA. The sequence can then be inferred directly form this *autoradiogram*; the shortest fragment will be found in the incubation mixture that contains the dideoxy analogue of the first base after the primer; the next shortest fragment will be found in the mixture that contains the dideoxy analogue of the second base after the primer; and so on for up to hundreds of bases.

DNA sequencing techniques were devised by molecular biologists and have been applied primarily to genetic problems with dramatic success, but to date the application of these techniques to taxonomic problems has been minimal. However, as the techniques are more widely applied, and a rapidly expanding library of sequences begins to include a number of homologous sequences form different organisms, taxonomists will undoubtedly come to depend increasingly on comparison of DNA base sequence to determine organismic relationships. Indeed it is probably not unrealistic to anticipate that in the near future the complete base sequence for the genomes of a number of microorganisms will be available, and microbial taxonomy may become on exact, quantitative science.

RNA Fingerprinting and Sequencing

Methods for sequencing RNA, comparable to those for sequencing DNA, have also been developed. They are, however, only beginning to receive wide application in microbial taxonomy. A number of tRNAs and 5S rRNAs have been sequenced but, because these are relatively small molecules, the amount of information they contain is fairly low. More useful for taxonomic purposes are sequences of larger molecules, such as the 16S or 18S rRNAs. Since sequencing these longer molecules was not feasible several years ago when Carl Woese initiated studies on the phylogeny of 16S RNA, alternative methods were devised. Since it was easy to sequence short RNA molecules, Woese digested 16S RNA with a specific endonuclease producing a variety of short oligonucleotides, which were separated and sequenced. This economical approach has allowed him and others to assemble catalogues of oligonucleotide sequences from hundreds of microorganisms; indeed it was the finding that the 16S rRNA sequences of methanogens, halophiles, and thermoacidophiles were unexpectedly divergent from those of other bacteria that allowed the recognition of the archaebacteria as a distinctive assemblage of bacteria.

The method devised by Woese and his collaborators is as follows: radioactive 16S rRNA (obtained by growing organisms in the presence of phosphate containing the radioactive isotope ^{32}P) is purified and digested with the endonuclease T_1. This endonuclease cleaves the RNA on the 3' side of every guanosine residue; hence, a mixture of oligonucleotides is obtained that range in size from a single guanosine residue at the 3' terminus. These are separated by two dimensional electrophoresis (the mixture is electrophoresed in one dimension in one buffer system, then in a second dimension in another buffer system) and their position determined by autoradiography. The labeled oligonucleotides can then be removed and sequenced. In practice, only those with a chain length of five residues or more contain sufficient taxonomic information to be worth cataloguing. Oligonucleotide catalogues of a number of organisms are analyzed with the aid of a computer program, which compares them in a pairwise fashion; results may be displayed as a dendrogram. Much of our current understanding of the broad outlines of bacterial phylogeny is based on the results obtained by this method.

In addition to the calculation of similarity coefficients among organisms, a detailed analysis of 16S rRNA oligonucleotide catalogues or complete sequences reveals that there are both quite variable and highly conserved regions of the molecule, presumably reflecting different degrees of functional constraint on different parts of the molecule. Separate comparison of the variable and conserved regions allows determination of both close and distant relationships. In addition, careful inspection of the conserved regions has revealed that there are sequences that are characteristic of major groups of bacteria, termed *signature sequences*. At great phylogenetic distances (very low S_J values) these signature sequences are important in allowing the determination of relationships.

More recently, rapid techniques for sequencing long molecules of RNA have been devised, and these are now used in preference to fingerprinting. To sequence a molecule of rRNA, bulk RNA is isolated form a culture (there is no need to purify it), and a host RNA primer is added that is homologous to one of the conserved regions of rRNA. Dideoxy sequencing can then be performed directly on the RNA template using the enzyme *reverse transcriptase*, a viral enzyme that produces a DNA complement to a RNA molecule. Comparison of complete sequences of 16S rRNA among organisms gives results identical to those obtained by comparing oligonucleotide catalogues, but with the advantage that more distant relationships can be determined with greater accuracy.

Bacterial Phylogeny

Complete sequence or oligonucleotide catalogues of 16S RNA have now been accumulated for more than 450 bacteria; analysis of these catalogues has allowed the recognition that the bacteria can be divided into several major groups based on the similarity of their ribosomal RNA. It is now generally agreed that the degree of sequence similarity of this molecule is a reflection of the phylogenetic distance among organisms. Because there are very tight constraints on the structure of rRNA (a consequence of the necessity for it to assume a precise secondary structure, and to interact with a variety of proteins to form a functional ribosome), the rate of change of the sequence of genes encoding it is much less than that of the bulk of the genome, and it is accordingly possible to determine relationships over vast evolutionary distances. Indeed, comparison of complete sequences of eubacterial and archaebacterial 16S and eukaryotic 18S rRNA shows that even at the greatest phylogenetic distances known among cellular organisms about 50 percent homology remains. Accordingly, this *rRNA functions as a molecular clock, and allows accurate determination of phylogenetic distance.*

It is, of course, possible that the phylogeny of the rRNA is not the same as the phylogeny of the rest of the genome; such a situation could occur as the result of genetic transfer among different organisms. If transfer of chromosomal material across large phylogenetic distance were common, the bacterial genome would be polyphyletic, and no single organismic phylogeny could be drawl. Although genetic transfer of plasmids across large phylogenetic distances is known, there is no evidence that transfer and integration of chromosomal material is common except among closely related organisms. Indeed, such genetic exchange would not be expected because homology between the transferred fragment and the endogenote would most probably be too low to permit recombination. The phylogeny of the 16S rRNA can thus be provisionally regarded as an organismic phylogeny. A critical test of this idea will come with the comparison of other sequences distributed over comparable phylogenetic distances and comparably conserved; possibilities include the genes for the proton-translocating ATPase, DNA polymerase, RNA polymerase, histonelike proteins, and ferredoxin.

Primary Division of Cellular Organisms

Comparison of oligonucleotide catalogues of eukaryotes, eubacteria, and archaebacteria shows very little detectable homology: S_J values across the boundaries of these groups are typically around 0.1, while within a group they are typically above 0.2. Analysis of complete sequences of 16S rRNA shows that there is less than 60 percent homology across group boundaries, but greater than 70 percent within a group.

Constituent Groups of Archaebacteria

Three major subgroups of archaebacteria can be distinguished on the basis of S_J values: the methanogen-halophile group: *Thermoplasma*; and the *Sulfolobus-Thermoproteus* group. The *Thermoplasma* group contains a single organism, while the other two groups are quite heterogeneous.

Constituent Groups of Eubacteria

Ten major groups of eubacteria are currently recognized. SJ values between groups are in most cases approximately 0.2–0.3, with the exception of the *Planctomyces* group, which is almost as distant from other eubacteria as it is from the archaebacteria.

Gram-positive bacteria

There are two major subgroups of the Gram-positive eubacteria: the *actinomycete group* or *high G + C group*, and the low G + C group. The low G + C groups is very heterogeneous one. It includes all the endospore-forming bacteria, the lactic acid bacteria, *Staphylococcus*, and the mollicutes. The genus *Clostridium*, defined on the basis of endospore formation and an anaerobic metabolism, is clearly polyphyletic.

Table 2.1. Constituent groups of *Eubacteria* based on rRNA sequence analysis

1. Gram-positive eubacteria
2. Purple bacteria—pseudomonad group
3. Spirochetes
4. *Bacteroids-Cytophaga Group*
5. Cyanobacteria
6. Green sulfur bacteria
7. Green nonsulfur bacteria
8. Sulfur-reducers and myxobacteria
9. Radioresistant micrococci
10. *Planctomyces* group.

Purple Bacteria and pseudomonads

Most of the Gram-negative bacteria fall within the confines of this very large and heterogeneous assemblage. Three subgroups can be discerned; the alpha and beta subgroups contain the purple nonsulfur bacteria, most of the spiral bacterial, and several rRNA homology groups of *Pseudomonas*; the gamma subgroup contains the purple sulfur bacteria, pseudomonads, spiral bacteria, and the entire group.

Spirochetes

All the helical bacteria motile by means of endoflagella fall into this phylogenetically coherent group. Two main groups of *Spirochaeta* are evident, one of which contains the treponemes. *Borrelia* and *Leptospira* are more distantly related (S_J values of 0.25 and 0.2, respectively, with the *Spirochaeta* group).

Bacteroides-cytophaga group

The Gram-negative anaerobic rods (*Bacteroides*) form a coherent phylogenetic group, with intra-groups S_J values above 0.45. These bacteria are distantly related to a heterogeneous group that contains the sheathed organisms *Haliscomenobacter*, *Flavobacterium*, and the gliding organisms *Cytophaga*, *Flexibacter*, *Sporocytophaga*, and *Saprospira*.

Cyanobacteria

The cyanobacteria form a coherent phylogenetic group that contains *Prochloron* and, at greater distances (S_J values of 0.25–0.3) the chloroplasts of red and green algae, and green plants.

Green sulfur bacteria and the green nonsulfur bacteria

The green photosynthetic bacteria comprise two distinct groups, very distantly related to each other (S_J values of about 0.18).

Sulfur reducers and myxobacteria

The anaerobic sulfur-and sulfate-reducing bacteria, the fruiting myxobacteria, and the bdellovibrios form a distinct cluster, although the relationships among the three subgroups are quite distant.

Radioresistant micrococci

Several species of the Gram-positive genus *Micrococcus* have long been known to be a typical: they are characterized by a high level of resistance to radiation, and an unusual cell wall. The finding that they are phylogenetically very distant from the other micrococci, constituting an isolated group, has led to the suggestion that they be renamed *Deinococcus*.

Planctomyces group

These budding bacteria constitute a heterogeneous group that shows very little detectable homology to either the eubacteria or the archaebacteria, although they are slightly closer to the eubacteria. Their lipids and their insensitivity to diphtheria toxin also suggest that they are a group of eubacteria. However, they are the only walled eubacteria to lack peptidoglycan.

Taxonomic Implications of Bacterial Phylogeny

Until very recently, firm phylogenetic conclusions could not be drawn regarding bacteria. The extent of morphological variation among the bacteria is quite narrow, and consequently the fossil record, in addition to being fragmentary, is uninformative about the details of bacterial evolution. Conclusions based on phenotypic similarity rested on assumption less reasonable than those that underlie phylogenies based on molecular evidence; furthermore, different conclusions resulted form approaches based on different phenotypic characteristics. Consequently, bacterial taxonomists largely abandoned the goal of a phylogenetic taxonomy and concentrated on devising systems that were useful and stable, based on phenotypic characteristics that often had no intrinsic importance.

The rapid expansion in recent years of our knowledge of the molecular homologies among bacteria, based on the techniques of nucleic acid reassociation, fingerprinting, and sequencing, has for the first time offered the opportunity for a reliable and objective phylogenetic taxonomy of the bacteria (provided, of course, that additional comparative studies confirm that the bacterial genome is largely monophyletic). Bacterial taxonomists are thus faced with a difficult choice. A decision to attempt to formulate bacterial systematics in phylogenetic terms would be in accord with the practice of other organismic taxonomies, and would establish a biological rationale for bacteriological systematics. However, extensive, confusing, and divisive nomenclatorial changes would be necessary; for instance inspection shows that many genera would have to be subdivided, some into as many as six genera. In many cases, the criteria for generic assignment of an organism include no easily determinable phenotypic traits, and depend solely on the molecular evidence, the routine determination of which is common in only a few laboratories in the world.

For these reasons, a change to a phylogenetic system of taxonomy will occur slowly, if at all. The taxonomy presented in the following chapters is accordingly in marked variance with the phylogenies presented in this chapter. Also, above the generic level, the practice established in previous edition of this book is continued: group of genera united by common physiological or morphological traits are grouped together, regardless of whether the molecular evidence indicates relationships.

3

Structure of Bacteria

Molecular Structure of Flagella and Pili

Although they differ both in function and in gross form, the two classes of filiform bacterial surface appendages, flagella and pili, share many common structural features. Both originate from the cell membrane and extend outward through the all to a distance that may be as much as 10 times the diameter of the cell. The external part of these organelles can be detached from the cell by mechanical means (e.g., shearing in a blender), and subsequently isolated and purified. The filaments of flagella and pili are made up of specific proteins, known as *flagellins* and *pilins*. The protein subunits (monomers) can be prepared by treatment of the isolated flagella or pili with heat or acid; they are of relatively low molecular weight (17,000 to 40,000). Studies of isolated flagella and pili by electron microscopy and X-ray diffraction have shown that the protein monomers are assembled in helical chains, wound around a central hollow core. The structure of the filament is consequently a reflection of the properties of the specific type of protein subunit from which it is built; it is determined by the size of the subunit and by the number and pitch of the helical chains into which they aggregate. The flagella of different bacteria differ slightly both in diameter (12 to 18 mm), and in form (i.e. height and wavelength of the helical curvature). Different types of pili differ greatly in width (4 to 35 nm). These minor variations within each class of organelle evidently reflect differences in the assembly properties of different flagellins and pilins. It has been shown that a single mutational change in the amino acid sequence of flagellin can cause a change in the height and wavelength of the flagellum formed from it.

Some bacteria bear flagella that are notably thickened. These *sheathed flagella* are surrounded by an extension of the cytoplasmic membrane. Under certain conditions of growth, some bacteria produce polar sheathed flagella along with many peritrichously arranged unsheathed flagella such flagellation is termed *mixed flagellation*. The probable ultrastructures of two filaments that have been studied in some detail, the flagellum of *Salmonella typhimurium* and type I pilus of *Escherichia coli*, are shown schematically in Fig. 3.1.

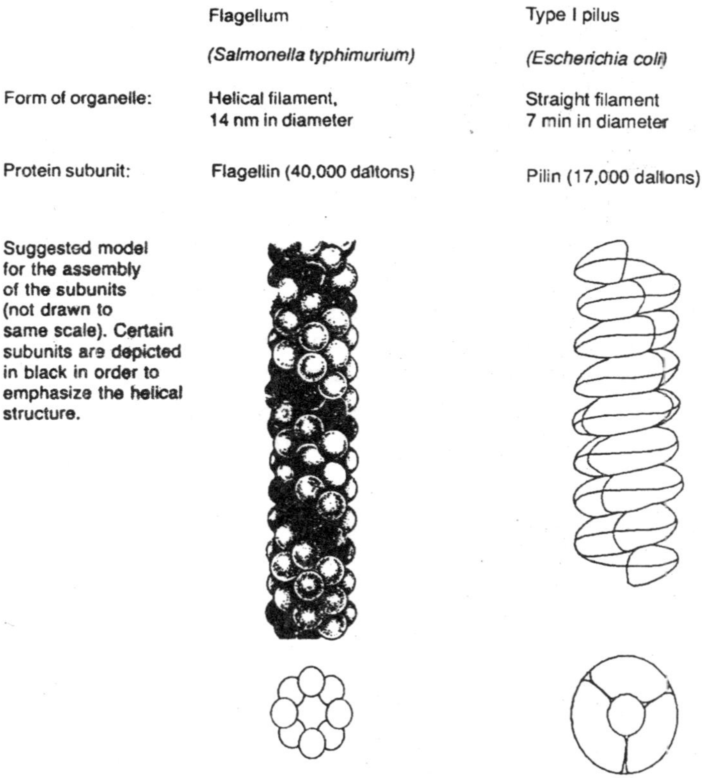

Fig. 3.1. Models showing the probable helical arrangement of the protein subunits of a bacterial flagellum (left) and of a pilus (right).

One function of type I pili is adhesion to surfaces including those of eukaryotic cells, and hence thee pili are sometimes essential to the first step in infection. In these cases the tip of the pilus attaches to specific receptor sites on the surface of the eukaryotic cell. Another type of pili, termed *sex pili*, which are morphologically similar to type I pili but which are composed of different pilin monomers, are synthesized by cells that contain plasmids that determine the cell's capacity to carry out conjugative genetic exchange with other cells. These pili also play an attachment role, which in this case is the first step in the process of conjugation.

Basal Structure of the Flagellum

The removal of flagella and pili by mechanical shearing breaks off these organelles near the cell surface; consequently it does not reveal their basal structures. However, by more gentle methods of cell breakage (osmotic lysis, detergent treatment), it is possible to isolate flagella with their basal structures intact.

The entire flagellar apparatus is made up of three distinct regions. The outermost region is the helical flagellar filament, of constant width, made up of flagellin. Near the cell surface this is attached to a slightly wider *hook*, about 45 nm long, made up of a different kind of protein, which is in turn attached to a *basal body*, located entirely within the cell envelope.

The basal body consists of a small central rod, inserted into a system of rings. In Gram-negative bacteria the basal body typically bears two pairs of rings. The outer pair (L and P rings) are situated

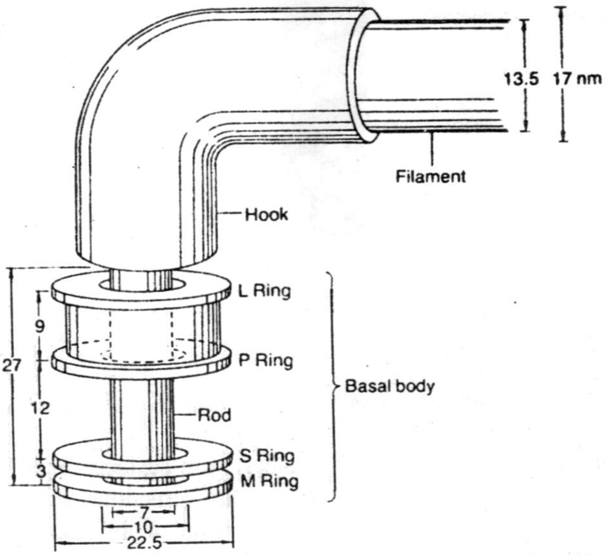

Fig. 3.2. Diagrammatic model of the basal end of the flagellum of E. coli, based on electron micrographs of the isolated organelle.

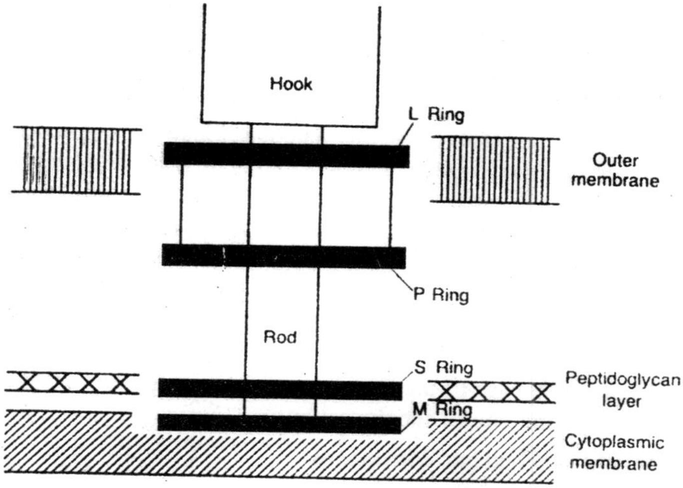

Fig. 3.3. A model showing the possible topological relations between the basal structure ofthe flagellum and the outer cell layers of E. coli. Dimensions in nanometers.

at the level of the outer membrane; apparently, their function is to serve as bushings for the insertion of the body through this layer. The inner pair (S and M rings) are located near the level of the cell membrane; the M ring is embedded either in it, or just below it, while the S ring lies just above, possibly attached to the inner surface of the peptidoglycan layer. On flagella of Gram-positive bacteria, only the lower (S and M) rings are present; apparently the upper pair is not required to support the rod as it passes through the relatively thick and homogeneous Gram-positive wall. This difference is significant, since it implies that only the S and M rings are essential for flagellar function.

Synthesis of the Flagellar Filament

Under appropriate physicochemical conditions, flagellin monomers can reassemble *in vitro* to produce filaments apparently identical with the flagella from which they are derived. This process of assembly requires the presence of seed structures: short flagellar fragments to which the flagellin molecules can attach. Such flagellar fragments possess a structural polarity; electron microscopy shows that one end is rounded and the other is indented. By marking the seed fragments with antiflagellar antibody, it can be demonstrated that flagellin sub units add only to the indented end. To which end of the intact flagellum, attached to the bacterial cells, does the indented end correspond? This question has been answered by an ingenious experiment on the growth of the flagella of *Salmonella typhimurium*. Addition of the amino acid analogue p-fluorophenylalanine to the growth medium causes *S. typhimurium* to synthesize abnormal ("curly") flagella of shorter wavelength than normal. When normal cells are cultivated for two to three hours in the presence of this analogue, and the distribution of curly waves on their flagella is then examined, it is found that these always occur on the distal parts of the flagella, the basal parts retaining the normal wavelength. Hence, it follows that the elongation of the flagellar filament takes place by the addition of the flagellar filament takes place by the addition of new flagellin subunits to its tip. It is most improbable that flagellins can attain the tip of a growing flagellum by excretion and passage through the medium; they are probably synthesized at the base of the flagellum and move outward through its hollow core to the site of incorporation.

Mechanisms of Flagellar Movement

The role of flagella as agents of bacterial movement can be demonstrated by a very simple experiment. If cells are mechanically deflagellated by shearing, they become immotile. Regrowth of flagella is rapid, the normal number and length being restored in about one generation. During flagella regrowth the cells at first show only rotatory movement; translational movement begins after the flagella have attained a critical length.

The way by which bacterial flagella propel the cell has been discussed for many decades, but experimental evidence in support of a specific mechanisms has been obtained only recently. This evidence, discussed below, indicated that flagella are semirigid helical rotors, each of which spins, either clockwise or counterclockwise, around its long axis. Movement is imparted to the organelle at its base by a flagellar "motor." It has been suggested that the motor operates by causing the S and M rings to rotate relative to each others. As already discussed, the M ring is situated in or just below the cell membrane. In order for relative rotation of the two rings to move the cell, it is essential that the S ring also be inserted into a structural fabric; most likely, it is attached to the inner surface of the cell wall. Such an attachment could account for the fact that complete removal of the bacterial cell wall by lysozyme treatment makes the protoplast immotile, even though it still bears intact flagella.

The cell evidently has the capacity to vary both the speed and the direction of rotation, as well as the frequency of stops and starts. Peritrichously flagellated bacteria swim in straight lines over moderate distances, these runs being interrupted periodically by abrupt, random changes of direction, known as *tumbles*. Recent observations suggest that smooth swimming of *E. coli* in a fixed direction is mediated by steady rotation of the flagella in a counterclockwise direction (viewing the flagellum along its rotational axis in the direction toward the cell). Tumbles are caused by a reversal of flagellar rotation, to the clockwise sense.

In spirilla, the multitrichous polar tufts of flagella are sufficiently thick to be visualized by phase contrast microscopy. During steady swimming the two polar tufts both rotate in the same sense. Spirilla never tumble; instead, changes in the direction of movement are brought about a reversal of the sense of rotation of the flagellar tufts, and they result in an exact reversal, by 180°, of the path previously followed. The mechanochemical basis for the operation of the flagellar motor is not known. However, there is good evidence that it is dependent on the transmembrane protonmotive force—not on the intracellular concentration of ATP.

Chemotactic Behaviour of Motile Bacteria

The cells of a suspension of flagellated bacteria are normally in a state of continuous but random active movement. However, if certain chemical gradients are imposed on the population, the cells migrate to and accumulate in that part of the gradient that provides an optimal concentration of the chemical. Some of the substances to which they can respond (for the most part, nutrients) act as *attractants*, in the sense that the cells will accumulate in the more concentrated region of the gradient. Others (for the most part, toxic substances) act as *repellants*, in the sense that the cells avoid regions of high concentration, and accumulate in that part of the gradient in which the concentration is lowest. This behaviour is known as *chemotaxis*. The specific substances that can elicit tactic responses differ from different bacteria; a particular chemical spectrum is characteristic for each species.

The system by which bacteria sense the concentration of attractants and repellants is far from completely understood but some key components have been discovered. Certain proteins, termed *binding proteins*, that participate in the entry of certain substrates into the cell by binding tightly and specifically to these substrates also serve as receptors for attractant and repellant signals. When an attractant or repellant binds to one of these proteins, a sequence of events involving, among other things, the methylation or demethylation of certain bacterial proteins transmits the signal to the flagellar motor.

Molecular oxygen elicits so-called *aerotactic* responses in most motile bacteria. Aerotactic patterns of accumulation can be readily observed in wet mounts in which an oxygen gradient is established by diffusion from the edges of the cover slip. Most strict aerobes accumulate close to the edge of the cover slip; however, the spirilla, which are microaerophils, accumulate in a narrow band some distance from the edge. For motile strict anaerobes, oxygen is a repellant; they accumulate in the center of the wet mount where the oxygen concentration is lowest.

Detailed studies of the tactic responses of *Escherichia coli* to organic compounds (sugars and amino acids) have shown that by no means all compounds that can serve as nutrients (energy sources) are attractants. For example, the disaccharide maltose is an attractant; the disaccharide lactose (also a good substrate) is not, although its split products (glucose and galactose) are. Conversely, the amino acid serine is a powerful attractant, although pyruvate, the first product formed from it, is not. Compounds that act as attractants are not necessarily metabolizable; for example, D-fucose, a nonmetabolizable analogue of D-galactose, is almost as good an attractant as galactose. Systematic studies on the responses of *E. coli* to sugars and amino acids have involved the use of mutants blocked either in the metabolism of attractants or in the ability to detect them, and have also involved the examination of competition between different attractants. These studies have led to the identification of 11 different chemoreceptors, each capable of eliciting a tactic response to certain specific compounds in the direction of swimming), which occurs every time they penetrate the sharp gradient of light intensity that separated the brightly illuminated spot from the surrounding dim area. As this observation suggests, purple bacteria and indeed all polarly flagellated bacteria differ from peritrichously flagellated bacteria in the manner by which

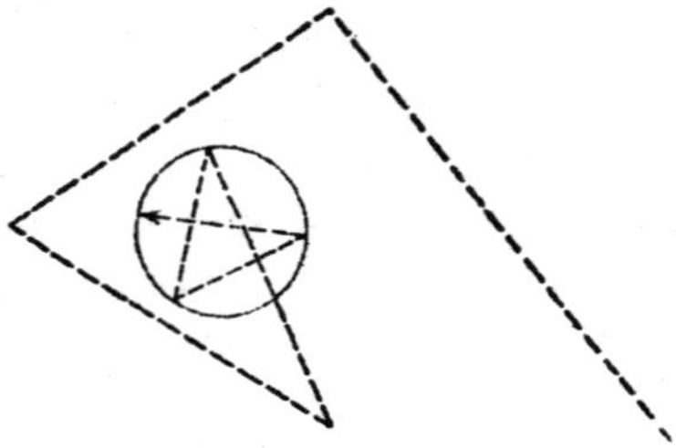

Fig. 3.4. A diagrammatic illustration of the mechanism by which phototactic bacteria accumulate within an illuminated area.

they change direction during a tactic responses. Rather than changing their direction by tumbling, they abruptly reverse their direction of swimming.

If a wet mount of motile purple bacteria is illuminated not with white light but with a spectrum produced by focusing light that has passed through a prism on the preparation, the bacteria rapidly accumulate in a series of bands corresponding to the principal absorption bands of their photosynthetic pigment system. A careful quantitative study of the relative effectiveness of different wavelengths in mediating phototaxis has shown that the action spectrum for phototaxis by purple bacteria corresponds exactly to the action spectrum for the performance of photosynthesis.

SPECIAL PROKARYOTIC ORGANELLES

As already mentioned, most prokaryotes do not form intracellular organelles bounded by unit membranes; the only possible exceptions are the thylakoids, which house the photosynthetic apparatus of cyanobacteria. However, three classes of prokaryotic organelles are bounded by *nonunit membranes*, made up of protein, at least in part, in all cases. These are *gas vesicles*, *chlorosomes*, and *carboxysomes* (*polyhedral bodies*).

Gas Vesicles and Gas Vacuoles

Since most cells have a slightly higher density than water, they tend to sink in an aqueous medium at a rate that is a function of cell size. This effect can be counteracted by swimming against the gravitational pull. However, many aquatic prokaryotes have developed another device to counteract gravitational pull: their cells contain gas-filled structures known as *gas vacuoles*. By light microscopy, gas vacuoles appear densely refractile, and have irregular contour. If such cells are subjected to a sudden, sharp increase in hydrostatic pressure, the gas vacuoles collapse, and the cells simultaneously lose their buoyancy and become much less refractile. Electron microscopy shows that gas vacuoles are compound organelles made up of a variable number of individual *gas vesicles*. Each gas vesicle is a hollow cylinder, 75 nm in diameter with conical ends, and from 200 to 1000 nm in length. The vesicle is bounded by a layer of protein 2 nm thick and is banded by regular rows of subunits ("ribs) running at right angles to its long axis.

Because the membrane is freely permeable to all common gases, the vesicles can neither tore nor accumulated gas. The composition and pressure of the gas in the vesicle are therefore functions of the dissolved gases in the surrounding medium. Water is excluded from the interior of the vesicles in the course of their formation and growth. This conclusion is confirmed by the observation that after pressure collapse, the vesicles do not recover; the cell can reacquire gas-filled vesicles only by the de novo synthesis of these structures.

Two factors are probably important in maintaining a gas phase within vesicles. One is the structural rigidity of the enclosing protein membrane, which must be able to resist the various forms of pressure that normally act upon it. The other is the chemical character of the membrane protein. Its outer surface is clearly hydrophilic, being readily wettable, but the inner wall is strongly hydrophobic.

Gas vacuoles occur in a wide variety of prokaryotes including archaebacteria, purple bacteria, cyanobacteria, green bacteria, and various groups of chemoheterotrophic eubacteria. The only common denominator of all these organisms is ecological: they occur in aquatic habitats. There can be little doubt, accordingly, that the function of gas vacuoles is to enable their possessors to regulate the buoyancy of the cell in order to occupy a position in the water column that is optimal for their metabolic activity with respect to light intensity, dissolved oxygen concentration, or the concentration of other nutrients.

Chlorosomes

In one group of photosynthetic prokaryotes, the green bacteria, part of the photosynthetic apparatus has a distinctive intracellular location: the antenna pigments are housed in a series of cigar-shaped vesicles, arranged in an cortical layer that immediately underlies, but is physically distinct from, the cell membrane. These structures, detectable only by electron microscopy, are 50 nm wide and 100 to 150 nm long, being enclosed by a single-layered membrane 3 to 5 nm thick. The photosynthetic pigments are entirely contained within them; these are probably the sites of the photosynthetic apparatus. The green bacteria are, accordingly, unique among photosynthetic organisms by virtue of the fact that the photosynthetic apparatus is not integrated into a unit membrane system.

Carboxysomes (Polyhedral Bodies)

A number of photosynthetic bacteria (cyanobacteria, certain purple bacteria) and chemoautotrophic bacteria (nitrifying bacteria, thiobacilli) contain structures termed *polyhedral bodies*, 50 to 500 nm wide, with, polygonal profiles, which are surrounded by a monolayer membrane about 3.5 nm wide and which have been isolated; they contain most of the cellular content of ribulose bisphosphate carboxylase (carboxydismutase), the key enzyme is the fixation of CO_2 associated with the operation of the Calvin-Benson cycle. They have been termed carboxysomes, and evidently represent the principal site of CO_2 fixation in these autotrophic prokaryotes.

Magnetosomes

In 1975 a remarkable group of bacteria were described by R.P. Blakemore that are magnetotactic; i.e. when they are placed in a magnetic field as weak as 0.2 gauss they orient and swim towards one or another of the magnetic poles. The sensing organelles, termed magnetosomes, within these cells are uniformly shaped enveloped crystals of magnetite (Fe_3O_4), a ferrimagnetic mineral. These are often arranged in a string.

Isolation of Bacterial Nuclei

If the bacterial chromosome consisted merely of a huge circular DNA molecule, folded in a random manner, its orderly replication and segregation could not possibly occur; some organization at a higher level must therefore be imposed on the chromosome. The recent isolation of structures that appear to be intact bacterial nuclei has provided a few clues to the nature of this organization. The structures in question have been obtained by gentle lysis of lysozyme-treated bacteria with nonionic detergents in 1.0 M NaCl. In addition to DNA, they contain a substantial amount of RNA and protein. They are rapidly sedimentable and are of low viscosity (in contrast to unfolded DNA of high molecular weight, which is extremely viscous).

Depending on the conditions of lysis, two structures of this type are obtainable. Lysis at 25°C releases structures with sedimentation coefficients of 1,300 to 2,200 S, which appear to be folded chromosomes, associated with a considerable amount of RNA and protein. The protein is largely RNA polymerase; and the RNA is newly transcribed single RNA strands. RNase treatment causes a rapid increase in viscosity, which indicated that some of the associated RNA is responsible for holding the DNA into a compact form. The DNA in these structures is folded into a number (between 12 and 80) of supercoiled loops.

If lysis is conducted at a lower temperature (0 to 4°C), structures having considerably higher rates of sedimentation (3,000 to 4,000 S) are obtained. Electron microscopy shows that they consist of folded

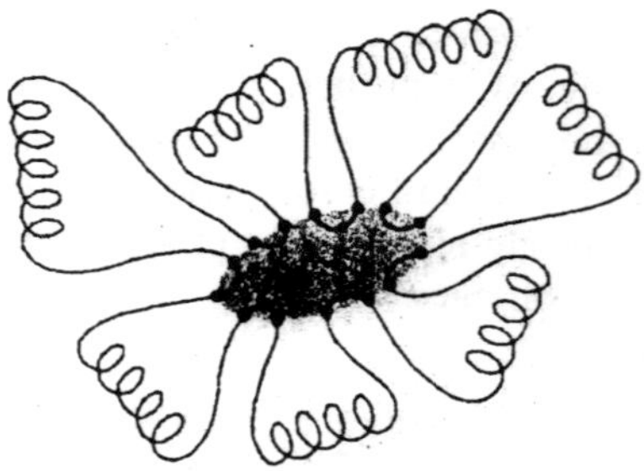

Fig. 3.5. A schematic drawing illustrating the proposed structure of the folded chromosome of E. coli.

chromosomes, attached to either one or two membrane fragments, from which they can be dissociated by gentle treatment. These observations show that binding of the chromosome to the membrane involves very weak forces. No folded chromosomes attached to membrane can be isolated from cells that have completed a round of DNA synthesis. Accordingly, it is possible that the resting chromosome may not be membrane-bound *in vivo*, membrane attachment taking place as a preliminary to the next round of replication.

4

FIELD BACTERIA

Bacteria belong to the class of organisms known as the *Schizomycetes (schizo, fission*, and *mycetes, fungi*). The organisms are single-celled and reproduce normally by transverse or binary fission.

The class Schizomycetes is divided into ten orders. The largest order is the *Eubacteriales;* it includes most of the common bacterial species.

Bacteria are typically unicellular plants, the cells being usually small, sometimes ultramicroscopic. They are frequently motile. By means of modern techniques, a true nucleus has been demonstrated in bacterial cells. Individual cells may be spherical or straight, curved or spiral rods. Cells may occur in regular or irregular masses, or even in cysts. Where they remain attached to each other after cell division, they may form chains or even definite trichomes. The latter may show some differentiation into holdfast cells and into motile or nonmotile reproductive cells. Some grow as branching mycelial threads whose diameter is not greater than that of ordinary bacterial cells, i.e., about 1μ. Some species produce pigments.

The true purple and green bacteria possess photosynthetic pigments much like or related to the true chlorophylls of higher plants. The phycocyanin found in blue-green algae does not occur in the Schizomycetes. Multiplication is typically by cell division. Endospores are formed by some species of Eubacteriales. Sporocysts are found in Myxobacteriales. Bacteria are free-living, saprophytic, parasitic, or even pathogenic. The latter types cause diseases of either plants or animals.

FILAMENT FORMATION

Cells that reproduce and divide in a normal manner may be induced to grow in filaments by changing the conditions of the medium. According to Webb, "the division of the bacterial cell follows a complex sequence, which in many respects, resembles that occurring in the cellular reproduction of higher forms. It is now known, for example, that bacterial cell division entails division of the nuclear element, division of the cytoplasm, secretion of new cell wall material, and the separation of the daughter cells.

Some or all of the events of this sequence are readily thrown out of balance, or even completely inhibited. Thus bacteria, particularly the rod-shaped organisms, may be induced to elongate into filaments by various treatments which apparently inhibit cell division but which do not inhibit growth. Such an effect is produced by various chemical substances, by sub-bacteriostatic concentration of certain antibacterial agents, as, for example, methyl violet, sulfonamides, *m*-cresol, penicillin, irradiation, and higher temperatures of incubation.

These changes in morphology induced by chemical substances are usually temporary, since reversion to normal form occurs promptly when the filamentous bacteria are subcultured in the absence of the inhibitory agents. Irradiation, on the other band, may give rise to a temporary or permanent induction of filamentous cells.

From observations such as these the concept has arisen that bacterial growth, in the sense of an irreversible increase in cell substance or volume, and cell division may be considered to some extent as separate and independent processes; at least, in so far as growth may occur either with or without the operation of the cell division mechanism."

Variation in the magnesium (Mg) content of the medium may exert a marked effect on cell division of some bacteria. In a Mg-deficient medium, Gram-positive rods grow in the form of long filaments. Such filaments revert to normal forms when transferred to the same medium supplemented with suitable concentrations of Mg. Filament formation is enhanced by the addition of zinc and cobalt. Inhibition of cell division occurs also in media supplemented with an excess of Mg.

Deibel et al. produced filamentous *Lactobacillus leichmannii* in the absence of vitamin B_{12}. Reversion to the normal cell form occurred on the addition of either vitamin B_{12} to a medium lacking the growth factor or of an excess of the desoxyriboside thymidine.

Shape of Bacteria

Bacteria exhibit three fundamental shapes: (1) spherical, (2) rod, and (3) spiral or curved rod. All bacteria exhibit pleomorphism in more or less degree under normal or other conditions, but a bacterial species is still generally associated with a definite cell form when grown on a standard medium under controlled conditions.

The spherical bacteria (singular, coccus; plural, cocci) divide in one, two, or three planes, producing pairs or chains, clusters, or packets of cells. Some are apparently perfect spheres; others are slightly elongated or ellipsoidal in shape.

The streptococci divide in only one plane. They grow normally in pairs or chains. Depending upon the species, the distal ends of each pair may be lancet-shaped, or flattened at the adjacent sides to resemble a coffee bean. The staphylococci divide in two planes, producing pairs, tetrads, or clusters of bacteria, the latter resembling bunches of grapes.

The sarcinae divide in three planes, producing regular packets. These are cubicle masses with one layer of bacteria atop another.

The rod forms also show considerable variation. A rod is usually considered to be a cylinder with the ends more or less rounded. Some rod forms are definitely ellipsoidal in shape. The ends of rods also show considerable variation. Some species are markedly rounded; others exhibit flat ends perpendicular to the sides. Gradations between these two forms may be seen.

Rods may show marked variation in their length/width ratio. Some rods are very long in comparison to their width; others are so short they may be confused with the spherical forms.

The shape of an organism may also vary depending upon certain environmental factors, such as temperature of incubation, age of the culture, concentration of the substrate, and composition of the medium. Bacteria usually exhibit their characteristic morphology in young cultures and on media possessing favorable conditions for growth.

Young cells are, in general, larger than old organisms of the same species. As a culture ages, the cells become progressively larger until a maximum is reached, after which the reverse effect occurs. Bacterial variations resulting from changes in age are only temporary; the original forms reappear when the organisms are transferred to fresh medium.

Size of Bacteria

Bacteria vary greatly in size according to the species. Some are so small they approach the limit of visibility when viewed with the light microscope. Others are so large they are almost visible with the normal eye. However, the sizes of the majority of bacteria occupy a range intermediate between these two extremes. Regardless of size, none can be clearly seen without the aid of a microscope.

A spherical form is measured by its diameter; a rod or spiral form by its length and width. Calculation of the length of a spiral organism by this method gives only the apparent length, not the true length. The true length may be computed by actually measuring the length of each turn of the spiral. Mathematical expressions have been formulated for making such computations.

The method employed for fixing and staining bacteria may make a difference in their size. The bacterial cell shrinks considerably during drying and fixing. This will vary somewhat depending upon the type of medium employed for their cultivation. Shrinkage generally averages about one-third of the length of the cell as compared to an unstained hanging-drop preparation. Young cells of *Bacillus megaterium* may shrink from 15 to 25 per cent when transferred from nutrient broth to the same medium containing sodium chloride in 2 M concentration.

Measurements show some variation depending upon the staining solution used and the method of application. In dried and fixed smears, the cell wall and slime layer do not stain with weakly staining dyes such as methylene blue but do stain with the intensely staining pararosaniline, new fuchsin, crystal violet, and methyl violet. The great majority of bacteria have been measured in fixed and stained preparations. In some instances dried, negatively stained smears have been used. Therefore, the method employed should be specified when measurements of bacteria are reported; otherwise the results will be of doubtful value. The unit for measuring bacteria is the micron. It is expressed by the symbol μ. It is 0.001 mm. or 0.0001 cm. A millimicron is 0.001 μ or 0.000001 mm. It is expressed by the symbol m_{μ}.

Some bacteria measure as large as 80 μ in length; others as small as 0.2 μ. However, the majority of the commonly encountered bacteria, including the disease producers, measure about 0.5 μ in diameter for the spherical cells and 0.5 by 2 to 3 μ for the rod forms. Bacteria producing spores are generally larger than the non-spore-producing species. The sizes of some common species in dried and stained smears are as follows: *Escherichia coli*, 0.5 by 1 to 3 *p.;* *Proteus vulgaris*, 0.5 to 1 by 1 to 3 μ; *Salmonella typhosa*, 0.6 to 0.7 by 2 to 3 μ; *Streptococcus lactis*, 0.5 to 1 μ in diameter; *S. pyogenes, 0.6* to 1 u in diameter; *Staphylococcus aureus*, 0.8 to 1 u in diameter; *Lactobacillus acidophilus*, 0.6 to 0.9 by 1.5 to 6 μ; *Bacillus subtilis* rods, 0.7 to 0.8 by 2 to 3 μ, spores, 0.6 to 0.9 by 1 to 1.5 μ; *B. megaterium rods*, 0.9 to 2.2 by 1 to 5 μ, spores, 1 to 1.2 by 1.5 to 2 μ; *B. anthracis rods*, 1 to 1.3 by 3 to 10 μ, spores, 0.8 to 1 by 1.3 to 1.5 μ.

The most commonly employed method for measuring bacteria is by means of an ocular micrometer. Measurements may also be made by using a camera-lucida attachment and drawing oculars, or by projecting the real image on a screen and measuring the bacteria.

The same factors that cause variations in the shape of bacteria also affect their size. With few exceptions, young cells are much larger than old or mature forms. Cells of *B. subtilis* from a 4-hr. culture measure five to seven times longer than cells from a 24-hr. culture. Variations in width are less pronounced. The organism *Corynebacterium diphtheriae* is a notable exception to the rule of decreasing cell size with age.

Variations in cell size with age are due to a variety of factors. The major causes appear to be changes in the environment with the accumulation of waste products. An increase in the osmotic pressure of the medium will also cause a decrease in cell size and may very well be the most important factor.

Bacterial Cell

Bacteria do not show the same morphological picture. Differences in structure exist between species. It is generally agreed that a bacterial cell consists of a compound membrane enclosing cytoplasm and nuclear material and often containing various granules, fat globules, and one or more vacuoles. In addition, some species contain resistant bodies known as spores, and some have one or more organs of locomotion called flagella.

The term protoplasm is used to indicate the thick viscous semifluid or almost jelly-like colorless, transparent material which makes up the essential substance of both the cell body and the nucleus, including the cytoplasmic membrane but not the cell wall. It contains a high percentage of water and holds fine granules in suspension.

Cytoplasmic Membrane

This membrane appears in young cells as an interfacial fluid film, becoming thicker and denser as surface-active material accumulates. It is finally converted into a firm structure. The membrane is believed to be composed mainly of lipide and protein. Polysaccharide has not been demonstrated as a component.

The membrane is acid in reaction because of its content of ribonucleic acid. It stains deeply with basic and neutral dyes over a wide range of pH. The membrane stains Gram-positive in Gram-positive bacteria and acid-fast in acid-fast organisms. It is a semipermeable membrane and is principally responsible for the Gram and acid-fast reactions. When a cell is plasmolyzed by immersion in a hypertonic solution, this membrane is drawn in with the cytoplasmic constituents. The thickness of the membrane varies even in a single cell. Measurements on a strain of *Bacillus cereus* at various stages of development ranged from 5 to 10 mμ in thickness.

Cell Wall

The cell wall is a more rigid structure and is responsible for the form of the bacterial body. It behaves as a selectively permeable membrane and apparently plays a fundamental role in the life activities of the cell.

The cell wall has a low affinity for dyes, which means that it is probably not stained in some of the usual staining procedures. It is lightly stained by certain basic dyes such as basic fuchsin and the methyl violets. Where deep staining of the wall is desired, the use of a mordant, such as tannic acid, is necessary. The mordant not only increases the affinity of the cell for dye, but it may increase the thickness of the wall.

The cell wall accounts for an average of about 20 per cent of the dry weight of bacteria and represents the major structural component. In thickness, it ranges from 10 to 23 mμ, depending upon the species.

According to Saltori, chemical analyses of cell walls have revealed differences in Gram-positive and Gram-negative bacteria. Cell walls of Gram-positive bacteria are lacking in aromatic and certain sulfur-containing amino acids, arginine, and proline. On the other hand, cell walls of Gram-negative bacteria show the presence of aromatic and sulfur-containing amino acids, arginine, and proline.

Gram-negative cell walls are generally richer in lipides than Grampositive bacteria.

Cell walls of a number of Gram-positive and negative bacteria contain the amino acid diaminopimelic acid.

Polysaccharides have been detected in both Gram-positive and Gramnegative bacteria. The polysaccharide is determined as reducing substances after acid hydrolysis. Some polysaccharides yielded only one reducing sugar; others yielded two or more sugars.

Gram-positive organisms gave rhamnose, galactose, and glucose; glucose only; rhamnose only; arabinose, galactose, and mannose. Gram-negative bacteria yielded galactose and glucose; galactose, glucose, mannose, and rhamnose.

In addition, all organisms studied contained an amino sugar or hexosamine. Generally, the walls of Gram-positive bacteria are richer in hexosamine than the Gram-negative forms.

Work pointed to the existence in Gram-positive bacteria of a common basal structure containing the following constituents: a hexosamine component comprising glucosamine and muramic acid and sometimes also galactosamind; a peptide component made up of alanine, glutamic acid, and either diaminopimelic acid or lysine with sometimes also glycine, aspartic acid, or serine; and usually a polysaccharide containing not more than four different sugar residues. Other substances may also be attached to the walls, as for example the protein antigens.

Smithies, Gibbons, and Bayley reported a relatively high nitrogen content in the walls of several halophilic bacteria which indicated that the cell material was predominantly protein. They contained only small amounts of lipides. The cell walls were lipoprotein.

Barkulis and Jones found that approximately one-third of streptococcal cell walls was made up of rhamnose and hexosamine. The remaining two-thirds was protein in nature.

For more information see Cummins and Harris (1956), Graziosi and Tecce (1957), Hayashi and Barkulis (1959), Kakutani (1957), Salton (1956), Strange (1959), Tanaka (1957), Tomcsik and Grace (1955), Trucco and Pardee (1958), Yoshida et al. (1956).

Capsules

Extracellular material of a slimy or gelatinous nature is formed by many bacteria, especially those producing mucoid growths. This material may remain firmly adherent as a discrete covering layer on each cell, or it may part freely from the cells. In the former case it is known as a capsule; in the latter, as free slime or gum.

Capsules and slime are believed to be distinct from the morphological and biochemical point of view. The capsule is a part of the cell, the slime a secretion. According to Klieneberger-Nobel, capsules are of definite shape, of more or less definite density throughout, and of definite outline, whereas slime envelopes are amorphous and can be drawn out into manifold structures, are most concentrated in the vicinity of the bacterial cells, and decrease in density with increasing distance from the cell.

Broth cultures of capsule-producing organisms are usually stringy in texture, and agar colonies exhibit a very moist, glistening surface which is described as mucoid. Capsule formation is dependent upon the composition of the medium but especially the variant phase of the organism. Some disease-producing organisms form large capsules in culture media rich in animal fluids. Others produce prominent dapsules when cultures are incubated at low temperatures (4 to 20°C.).

Chemical analyses of capsular material from a number of bacteria show wide differences in composition. For this reason it is impossible to make statements which apply to all bacteria. In some organisms the capsular material appears to be a glycoprotein; in others, a protein-polysaccharide complex; in still others, a polysaccharide framework with the spaces filled in by a larger amount of glutamyl polypeptide.

Capsular material is difficult to distinguish from those gums which flow away from the cells as they are formed. Organisms producing gums do so when grown in sugar solutions. Some organisms produce gums only in the presence of a specific sugar; others produce gums in the presence of any one of several sugars. In the absence of sugar, usually very little, if any, gum is formed. Organisms producing gums of this type are the cause of considerable losses in the sugar industry. The increased viscosity produced by the gum interferes with the filtration of the sugar solution.

The species commonly encountered in sugar-cane juice is *Leuconostoc esenteroides*. The cells are surrounded by a thick, gelatinous, colorless polysaccharide consisting of dextran (glucose polymer). The formation of gums is of common occurrence by soil bacteria. From 5 to 16 per cent of such forms have been shown to be capable of synthesizing gums from sugars.

When the cell wall is damaged, the protoplasm usually disintegrates. However, methods are available for removing the cell membranes without destroying the vital nature of protoplasm. The term protoplast is used to indicate living protoplasm exclusive of the cell membranes.

Action of lysozyme

Some bacteria are rapidly lysed or dissolved by the action of lysozyme. Weibull reported that lysozyme possessed a specific depolymerizing action on the cell wall and that this appeared to be the only portion of the cell that was affected by such treatment. As a result of the destruction of the wall, the protoplasts were liberated.

Protoplasmic structures are not very stable. Consequently, if protective agents were not employed, destruction of the walls was accompanied by a rapid lysis of the protoplasts, followed by the liberation of most of the cell protein and nucleic acid in soluble form. This could be prevented by employment of the enzyme in a 0.2 M solution of sucrose or cane sugar. After digestion of the cell walls, the living protoplasts rounded up into spheres.

Spiegelman, Aronson, and Fitz-James found that digestion of protoplasts of *Bacillus megaterium* led to the liberation of nuclear bodies of the protoplasts. Such bodies were collected by centrifugation for 5 min. at 10,000 × g.

Properties of protoplasts

The difficulty in handling and studying protoplasts is their extreme fragility and sensitivity to osmotic shock, shaking, centrifugation, and aeration. Removal of the cell wall does not change the structure and capabilities of the protoplasm. Permeability, respiration, and spore formation appear to be the same for protoplasts and intact cells. Also both can support the development of bacteriophages. Under special conditions the protoplasts grow and probably divide like intact cells. However, there is no evidence that protoplasts form colonies.) The metabolism of protoplasts and intact cells appear to be very similar but probably not identical.

Filterability of protoplasts

Sinkovics reported the spontaneous occurrence of units in aged cultures of *Escherichia coli* which conformed to the description of artificially induced bacterial protoplasts. The disintegration of cells from aged cultures was preceded by swelling of the cell and rupture of the rigid cell wall. Centrifugation of the culture gave a supernate which, after filtration, contained units capable of regeneration when placed in fresh medium. The smallest units capable of regeneration measured about 350 m1s in diameter. The units underwent fusion before cell-wall formation occurred.

The results supported the assumption that aged *E. coli* cultures could survive in the form of units having no cell walls and which, under adequate conditions, regenerated into vegetative forms.

Polysaccharide Structures

Polysaccharides occur (1) in cell walls, (2) extracellularly in capsules and gums, and (3) inside of bacterial cells. The first two have already been discussed.

Pennington (1949) revealed the presence of polysaccharides by treating bacteria with sodium metaperiodate followed by staining with sulfitedecolorized basic fuchsin. In *Bacillus cereus* the polysaccharide was concentrated in the cytoplasmic membrane as well as in the cell wall.

Selective staining of polysaccharide in the cell is said to depend upon the oxidizing action of periodate on such chemical configurations as α, β glycols and a-hydroxyketones. Polyaldehydes generated by this selective oxidation react with sulfite-decolorized fuchsin. Polysaccharide areas in the cell are colored red by the stain.

Nucleus

The question of the presence of a well-defined nucleus in bacteria has been the subject of investigations by bacteriologists almost from the beginning of bacteriology.

Some of the earlier cytologists maintained that bacteria were very primitive organisms devoid of nuclei and consisting simply of cytoplasm, granules, and vacuoles. This view was based on their failure to observe a nucleus in a bacterial cell. Others held the view that the nuclear material was present in a diffuse form throughout the cytoplasm. Still others believed that the whole cell should be regarded as a "naked nucleus," corresponding to the nucleus of higher organisms. The naked nucleus is regarded as a primitive form of living matter. Since bacteria have the structural and physiological attributes of true cells, this concept cannot apply to these organisms.

Much of the confusion was caused by the inadequacy of the staining procedures. By means of the HCl-Giemsa staining technique of

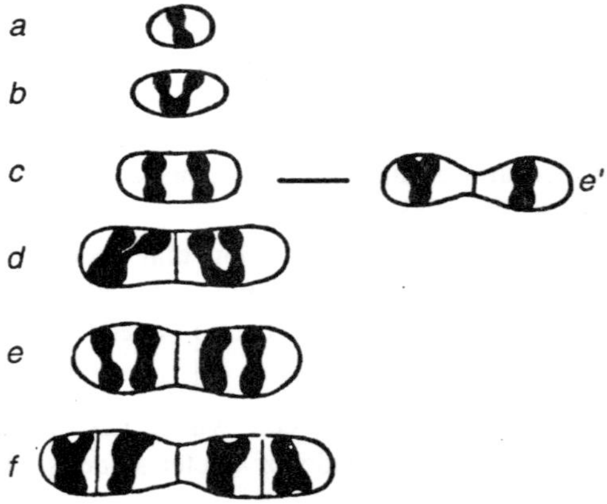

Fig. 4.1. Escherichia coli. Diagram of successive division stages of the chromatinic bodies from the beginning of the lag phase, after transfer to a fresh nutrient medium, to the first division of the growing organism. c—c′ and c—f are alternative modes of development, c—f being that most commonly followed.

Piekarski, many observations have been reported demonstrating the presence of chromatinic structures in bacteria.

Robinow prepared wet smears of *Escherichia coli*. Slides were fixed in osmic acid vapor, dried, and immersed in normal HCl for about 9 min. at 53 to 55°C., then washed and stained in 1:20 Giemsa solution for 10 to 60 min., depending on the staining properties of the specimen. The chromatinic structures in *E. coli* from old cultures were too small to be resolved accurately. After transfer to fresh medium the chromatinic structures increased in size and gave rise to short, often dumbbell-shaped rods or chromosomes, which multiplied by splitting lengthwise in a plane more or less parallel to the short axis of the cell. A single cell of *E. coli* contained one chromatinic body or one or two pairs of these representing primary and secondary division products.

The Smith technique consisted of fixing the smear in osmium tetroxide vapor, immersion in HCI, mordanting in dilute formaldehyde, and staining with aqueous basic fuchsin. The method was said to possess certain advantages over the procedure of Robinow.

Another cause of confusion in the recognition of nuclear structures in bacteria was the lack of appreciation for the ages of the cultures. At certain times nuclear structures cannot be seen. In general, most of the early observations of tidy, intelligible "nuclei" were made on preparations from very young cultures, whereas haphazardly scattered, unintelligible granules of chromatin were persistently observed in preparations made from cultures of the same bacteria beyond the logarithmic growth phase.

Recognition of the fact that the configuration of nuclear material in bacteria might change with age removed one of the chief causes of confusion. As Dobell said (1911), "My own belief is that the nucleus in bacteria may display not one but many forms during the whole life cycle. Many of the nuclear structures which have been shown to exist in these organisms should, I think, be regarded as temporary states rather than aspermanent conditions. The different results which have been reached by different workers when working, apparently, upon the same species, may to some extent find an explanation in this circumstance."

The chromatin bodies, according to Robinow, appear to have the following properties:

They are simple structures of relatively low density, not markedly basophilic but reacting positively in the Feulgen test. Normally they lie separately in the cytoplasm, and all those in one bacterium are homologous. Changes in the balance of ions in the cytoplasm may cause the aggregation of several chromatin bodies into a single continuous structure. This effect is reversible. Growth and division of chromatin bodies are attended by changes of form only, not by visible changes of texture. In the simplest type of body, that which in profile looks like a bar or dumbbell, division begins at one end and causes the successive appearance of V-, U-, and H-shaped phases. The chromatin structures of certain bacteria are netlike or spongelike, and their mode of division is not easily imagined.

Direct division of the chromatin bodies of *E. coli* has been demonstrated by Mason and Powelson in a series of remarkable phasecontrast photomicrographs. These observations were made on living bacteria. The nuclear areas in the dividing cells appeared to be as clearly defined as the areas in fixed, hydrolyzed, and stained cells.

Vacuoles. Vacuoles have been identified in young bacteria. They are cavities in the protoplasm and contain a fluid known as cell sap. As the cells approach maturity, some of the water-soluble reserve food materials manufactured by the cell dissolve in the vacuoles. Insoluble constituents precipitate out as cytoplasmic inclusion bodies.

Metachromatic Granules

The best-known inclusion bodies in bacterial cells are known as volutin or metachromatic granules. The granules are small in young cells and become larger with the age of the culture. They are believed to originate in the cytoplasm of young cells and to localize in the vacuoles of mature forms. The granules show a "Strong affinity for basic dyes, indicating that they are acid in character. They are usually considered to be a reserve source of food. Grula and Hartsell found the granules to be composed of metaphosphate or another form of

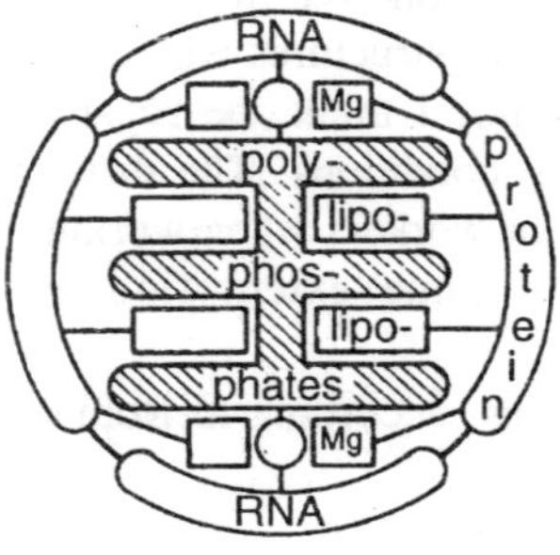

Fig. 4.2. Composite representation of a metachromatic granule.

inorganic phosphate, some fat, and possibly small amounts of protein. Their presence and size in cells were related to the phosphate concentration of the growth medium in the presence of an energy source and specific divalent ions (Mn and Zn). Older cells possessed larger granules. Their basophilic nature did not depend on either ribonucleic acid (RNA) or desoxyribonucleic acid.

On the other hand, Widra found metachromatic granules to contain protein-bound lipide, RNA, and polyphosphates.

Fat Globules

Bacteria are capable of storing fat in the form of globules. Fat globules may be demonstrated in 24-hr. cultures and usually reach a maximum in about 48 hr.

Some cells may contain only one large globule; others may show the presence of a number of small, scattered globules.

It is generally believed that fat is stored as reserve food material. Globules usually cannot be demonstrated in young, vigorously growing cells. As cells age and slow down in activity, fat globules appear in the cytoplasm and may be recognized by appropriate staining.

Motility

Bacterial motion is generally associated with the presence of organs of locomotion known as flagella (singular, flagellum). They were first observed in stained preparations by Cohn. The presence of flagella does not mean necessarily that the organisms are always motile, but it indicates a potential power to move.

Independent bacterial motion is a true movement of translation and must be distinguished from the quivering or back-and-forth motion exhibited by very small particles suspended in a liquid. This latter type of motion is called Brownian movement and is caused by the bombardment of the bacteria by the molecules of the suspending fluid.

Properties of flagella

Flagella are very delicate organs and easily detached from the cell. In the stained condition they are long, slender, undulating organs. They are directed backward to the direction of motion at an angle of about 45°. Reversal of direction occurs by swinging the flagella through an angle of about 90°. Turning movements take place by swinging the flagella forward on one side only. They propel the organism by a spiral or corkscrew motion.

The thickness of flagella varies from species to species. In *Proteus vulgaris* they measure about 12 mμ. This figure is considerably below the shortest wave length of visible light and explains why flagella cannot be seen in hanging-drop preparations or in smears stained by the usual simple procedures. When special staining methods are employed, sufficient dye becomes deposited on the flagella to make their diameters greater than the wave length of visible light. They may then be seen under a light microscope.

Chemistry of flagella

Flagella and bacterial bodies differ in composition. Flagella break up on boiling or when exposed to pH values below 4 or above 11. Their composition is largely protein, having a molecular weight of about 41,000. Weibull found the flagella of *P. vulgaris* to be composed of 98 per cent protein, traces of carbohydrate and fat, and no phosphorus. The protein contained only 14 known amino acids. It is an incomplete protein, lacking in some of the essential amino acids.

It is well established that the H antigens of bacteria are associated with the flagella and the 0 antigens with the bodies. Purified flagella are agglutinated by *H* antiserum but not by 0 antiserum; 0 antigens are not agglutinated by *H* antiserum. This is another indication that flagella and bacterial bodies differ in composition.

Origin of flagella

Some believe flagella originate from the cell wall; others believe they traverse the cell wall into the protoplasm. Flagella differ chemically both from the cell wall and the protoplasm. Two observations have been made relative to the site of origin of flagella. Electron micrographs by van Iterson and others show the flagella penetrating the faint outer zones and extending into the cytoplasm. If the outer zone is the cell wall, then the flagella have their origin in the cytoplasm.

Weibull showed that removal of the cell wall of some bacteria by means of lysozyme produces a spherical protoplast which still retains the flagella of the treated cell. The obvious conclusion is that flagella have their origin in some cell structure deeper than the cell wall.

Number and arrangement of flagella

The number and arrangement of flagella vary with different bacteria, but they are generally constant for each species. Some have only one flagellum; others have two or more flagella.

In rod-shaped cells the flagella arise either at one or both poles, or are distributed laterally with the poles being generally bare. In some species flagella are located both laterally and at the poles. A species may show considerable variation in the number and arrangement of flagella. Single *Alcaligenes* cultures may contain forms with a polar flagellum only; some with several lateral flagella; and some with both lateral and polar flagella. Sometimes a species may show cells which are flagellated in one environment and nonflagellated in another.

Leifson, Carhart, and Fulton reported the presence of four definite types of curvature in the flagella of *Proteus vulgaris*. Individual organisms may have more than one type of flagella, and individual flagella may have one or two types of curves.

Environmental factors, particularly pH, may change the curvature of the flagella on some strains, but not on all strains. In acid media the curly curvature tends to predominate; in alkaline media the normal predominates.

Organisms have been classified on the basis of the number and arrangement of flagella as follows:

Monotrichous—a single flagellum at one end of the cell.

Lophotrichous—two or more flagella at one end or both ends of the cell.

Amphitrichous—one flagellum at each end.

Peritrichous—flagella surrounding the cell.

Staining of flagella

The staining of flagella is a difficult technique, especially in the hands of the beginner. For this reason many methods have been proposed. Regardless of the method employed, the film must first be treated with a mordant to make the flagella take the stain heavily. Mordants consist usually of a mixture of tannic acid and some metallic salt. In some methods the mordant and stain are applied separately; in others they are combined in one solution.

Boltjes came to the following conclusions on the staining of bacterial flagella:

Four factors at least influence the results of staining-the skill of the investigator, the organism studied, the culture medium on which the organism was grown, and the staining method. Of these the first is perhaps the most and the last the least important. The importance of skill is shown by the repeated failure of students to stain flagella although their teacher has no difficulty in demonstrating flagella at the same time and with the same suspension. Further, one usually has success with a new formula only after a number of trials; first attempts to stain an unknown bacterium are often a failure. Another point is that as a rule flagella can be clearly seen only in a rather small part of the preparation. The different colours of the stained bacteria show clearly that during the staining process conditions are not everywhere alike, and since many flagella are torn off during drying, we need not wonder that in most cases only a few bacteria are successfully stained. Notwithstanding all this, it is certain that when one has had some experience with the staining technique, the results are so consistent that there will never be any confusion between bacteria with true polar flagella, such, for example, a, *Pseudomonas fluorescens* or *Vibrio comma,* and peritrichous bacteria like *Proteus mirabilis* and *Salmonella typhosa. I* therefore consider flagella staining to be a reliable procedure.

Photomicrographs of the arrangement of flagella on bacteria, according to Leifson.

Pijper Theory of Motility

In a series of investigations Pijper et al. questioned the belief that flagella are responsible for motility. He added methyl cellulose to a culture of a motile organism to increase the viscosity of the medium. This treatment decreased the motility of the cells. Under these conditions the cells exhibited a gyratory undulating movement like other aquatic creatures. He concluded that flagella were not organs of locomotion but only artifacts-useless appendages, polysaccharide twirls-the result, not the cause, of bacterial motility. To quote, "That motile bacteria always exhibit a gyratory undulating movement was confirmed by making a slow motion cinemicrographic film of fast-swimming bacteria in broth, and also by examining the same bacteria at lower temperatures, which reduced their speed.

This spirillar motion of bacteria is sufficient to propel them, and there is no need to invoke special motor organs like flagella. There is no evidence to show that the flagella-like appendages of bacteria act as motile organs-in fact all the evidence when critically examined points the other way. Analysis of the structure of bacteria excludes the possibility that tails, "flagella," or the thin wavy threads are live organs, or that they are in direct communication with the living parts of the cell. There is no evidence from either electron pictures or stained preparations that it is otherwise.

Not only does the visible gyratory undulating movement of motile bacteria satisfy all requirements for locomotion, but it is possible for bacteria grown under special conditions to swim in this fashion without showing tails or other supposed motor organs."

Notwithstanding the findings and conclusions of Pijper, evidence at present appears to be overwhelmingly in favor of flagella as organs of locomotion. As has already been stated, Weibull reported the flagella of *P. vulgaris* to be composed of 98 per cent protein. This contradicts Pijper's statement that flagella are formed from the carbohydrate slime layer that is peeled off into a number of thin, wavy threads.

Several investigators, among them Labaw and Mosley, by means of electron microscopy, demonstrated the presence on *Brucella bronchiseptica* of uniform flagella having an external contour of a counterclockwise or left-handed triple helix. The average periodicity along the length of the flagella was 19 mμ, with an average diameter of 13.9 mμ.

Motion of Colonies

Several organisms have been described which exhibit colonial motility when grown on a solid medium.

Shinn prepared lapse-time motion pictures of individual colonies of *Bacillus alvei* grown on agar plates and measured their velocities. The linear motions of colonies measuring 0.2 to 0.5 mm. in diameter averaged about 14 mm. per hr. Comparing this figure with the speed of individual cells of other species of motile bacteria gave the following results:

Salmonella typhosa	65 mm. per hr.
Bacillus megaterium	27 mm. per hr.
B. alvei (colonies)	14 mm. per hr.

The colonies exhibited not only linear motion but also a slow rotary movement. The direction of rotation of 200 to 300 colonies observed was counterclockwise, with the exception of two colonies in which it was clockwise.

Turner and Eales reported that the rotation of an aerobe occurred very early during growth. The cells segregated in small groups and aligned themselves concentrically around a common center to form disk-like plaques one or a few cells thick. The rate of rotation was greater in smaller groups. As multiplication continued, successive layers were gradually built up in terrace fashion and th colony grew in height. Teh colonies then began to migrate. When a colony migrated, it left a peculiar "track" on the surface of the agar. A small number of cells were left behind, mostly at the edges of the track, which formed two parallel lines separated by the width of the moving colony.

Typical migrating colonies pursued curved or spiral paths which were often very elaborate and of relatively great length, even 2 or 3 cm.. The direction of rotation was either clockwise or counterclockwise.

After wandering for a variable distance, a colony approached the center of its spiral path with rapidly shortening radius, ceased to migrate, began to rotate around its center, lost its elongated shape, and increased in size to several times the width of the track at the end of which it was formed.

Endospores. Endospores are bodies produced within the cells of a considerable number of bacterial species. They are more resistant to unfavorable environmental conditions, such as heat, cold, desiccation, osmosis, and chemicals, than the vegetative cells producing them. However, it is debatable if such extreme conditions -actually occur in nature. For instance, the resistance of spores to high temperatures is a laboratory phenomenon and probably never occurs in a natural environment.

The bulk of evidence indicates the existence of a close relationship between spore formation and the exhaustion of nutrients essential for continued vegetative growth. Sporulation is a defense mechanism to protect the cell when the occasion arises.

Spore formation is limited almost entirely to two genera of rod-shaped bacteria: *Bacillus* (aerobic or facultatively anaerobic), and *Clostridium* (anaerobic or aerotolerant). With one possible exception, the common spherical bacteria do not sporulate. Some spore-bearing species can be made to lose their ability to produce spores. When the ability to produce spores is once lost, it is seldom regained. SporMation is not a process to increase bacterial numbers because a cell rarely produces more than one spore.

Morphology of spores

Spores may be spherical, ellipsoidal, or cylindrical in shape. The position of the spore in a cell may be central, subterminal, or terminal. A fully grown spore may have a diameter greater than that of the vegetative cell. This causes a bulging of the cell. The resulting forms are known as clostridium if central, and plectridium if terminal. As a rule, each species has its own characteristic size, shape, and position of the spore, but this is subject to variation under different environmental conditions.

Franklin and Bradley, by means of electron microscopy of carbon replicas, reported that the spores of a majority of species of *Bacillus* and *Clostridium* are readily distinguished by surface patterns. The surfaces may be smooth or ribbed, with the ribs usually longitudinal.

The sculpturing consists of a single endless ridge in the form of two loops, similar to the marking on a tennis ball, together with two

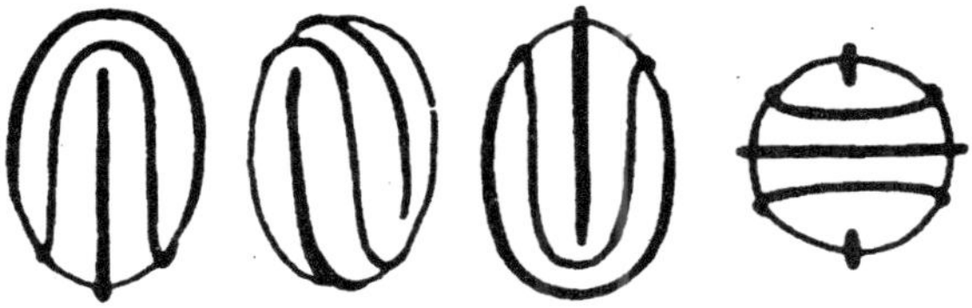

Fig. 4.3. Surface structure of a spore of Bacillus polymyxa. From left to right: side view; same rotated a quarter turn from right to left; same rotated a further quarter turn; view of a pole.

other separate ridges terminating within the loops. An electron micrograph of ultrathin sections of such spores, by van den Hooff and Aninga. The spore coat consists of an outer and an inner layer separated by a space. The outer layer is sometimes called the exine and the inner layer the intine. The intine faintly follows the surface relief. The central core is separated from the intine by a regular nonosmophilic space. A peripheral spot may be observed in the core which probably represents nuclear material.

Parasporal bodies

When sporulation of *Bacillus laterosporus is* complete, the spores are cradled in canoe-shaped bodies. According to Hannay (1957): "On sporulation the slender vegetative rods swell and form larger spindle-shaped cells in which the spores are formed. When the spores mature they lie in a lateral position cradled in canoe-shaped parasporal bodies which are highly basophilic and can be differentiated from the surrounding vegetative cell cytoplasm with dilute basic dyes. On completion of sporulation the vegetative cell protoplasm and the cell wall lyse, leaving the spore cradled in its parasporal body. This attachment continues indefinitely on the usual culture medium and even persists after the spores have germinated. In thin sections of sporing cells the bodies are differentiated from the cell protoplasm by differences in structure. Whereas the protoplasm has a granular appearance, in both longitudinal and cross-sections the parasporal body comprises electrondense lamellae running parallel with the membranes of the spore coat and less electrondense material in the interstices of the lamellae. The inner surface of the body is contiguous with that of the spore coat as if it were part of the spore, rather than a separate body attached to the spore. The staining reactions of the parasporal body are not consistent with those of any substance descred in bacteria."

Composition of spores

Ross and Billing, by means of refractive index measurements on spores and vegetative cells of *B. cereus, B. cereus* var. *mycoides,* and

B. megaterium, found the values to be very high and comparable with that of dehydrated protein. This suggested that they contained much less water than the vegetative cells.

Strange and Dark demonstrated the presence of a hexosamine containing peptide in the spore coats of *B. megaterium* and *B. subtilis.* The breakdown of an insoluble peptide complex might well be one of the first steps of the germination process. It was believed that the release of the hexosamine-amino acid complex was the result of the action of lysozyme present in the spores.

Enzymes of spores

The presence of enzyme systems in *Bacillus* spores has been reported. Some of these are an inorganic pyrophosphatase that requires manganese for activation, adenine ribosidase that hydrolyzes adenosine, an enzyme that functions possibly to lyse the sporangium and free the spore during germination, highly active alanine racemase that catalyzes the conversion of L-alanine to n-alanine, several glucose dehydrogenases, and an aldolase.

Sporulation process

Conditions necessary for sporulation in one species do not necessarily apply to another. The subject appears to be in such a state of confusion that it is impossible to discuss sporulation in terms of generalities.

The conditions which have been reported as favoring sporulation include addition of salts of metals such as manganese, chromium, nickel, etc., to the medium; shaking a culture of vegetative cells of sporing aerobes with distilled water at 37°C.; addition of tomato juice to a medium; incubating the cultures at an appropriate temperature; addition of calcium carbonate to a carbohydrate medium to prevent excessive accumulation of acid, and to maintain the pH at 5.5 or above; the necessity of ygen; addition to the medium of certain amino acids; etc.

Germination of spores

With the exception of some constituents such as high concentrations of calcium, dipicolinic acid, and in *Bacillus sphaericus,* a, E-diaminopimelic acid, spores are similar to the vegetative cells in composition.

When a spore prepares itself for germination, it loses its refractility, which coincides with an imbibition of water. This stage is associated with a loss in heat resistance, stainability, and dry weight. Later the spore coat breaks, followed by the emergence from the spore case of a new germ cell which eventually matures into a vegetative cell.

Spore germination has been defined in various ways. According to Campbell, "Spore germination may be regarded as the change from a heat resistant spore to a heat labile entity which may not necessarily be a true vegetative cell." Later development, leading eventually to the formation of a mature vegetative cell, is called outgrowth.

Some conditions which stimulate germination are as follows: (1) Treatment at 90 to 100°C. for 1 to 2 min. stimulates germination. (2) Spores which fail to germinate overcome this dormancy when activated by heat. (3) The use of certain agents such as alanine, glucose, and adenosine stimulates spore germination in most sporing species. In some species other amino acids may be substituted for the alanine. The same applies to glucose. (4) Yeast extract and mixtures of vitamin-free amino acids have also been shown to stimulate germination.

Spores germinate in a variety of ways. There is a considerable degree of constancy in the method of spore germination for each species. Lamanna classified the modes of germination as follows:

I. Spore germination by shedding of spore coat. Characteristics of this method. are
   A. Spore does not expand greatly in volume previous to the germ cell breaking through the spore coat. The limit of volume increase of the spore may be considered to be twice its original volume.
   B. Spore coat does not lose all its refractive property previous to germination.
   C. After the second division of the germ cell, giving a chain of three organisms, the original spore coat, remaining attached to the cells, is visible for a long time after germination.
      1. Equatorial germination.
      2. Polar germination
      3. Comma-shaped expansion.

II. Spore germination by absorption of the spore coat. Characteristics of this method are
   A. The spore expands greatly during germination. A tripling or greater increase of the original volume occurs.
   B. The spore loses its characteristic refractiveness during germination, so that it is difficult to say when the sore has disappeared and the germ cell appeared.

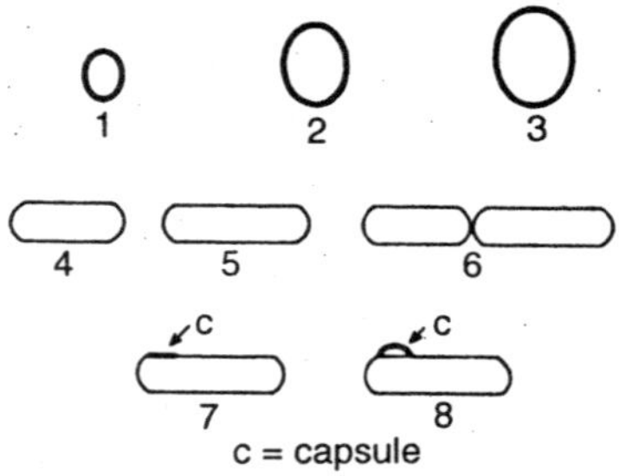

Fig. 4.4. Spore germination by absorption.

C. After the second division of the germ cell even if a thin capsule originally remains, all traces of the spore coat are gone.

Some strains germinating by absorption regularly show a thin capsule remaining about one end of the growing cell. This would appear as a polar germination. In other cases, equatorial capsules are seen. Yet, in all instances, the spore is considered to germinate by absorption inasmuch as the three characteristics of the method are still adhered to.

5

Soil Bacteria

It is generally known that soils are derived from weathered rock. However, a mass of weathered rock does not make a soil without the intervention of living processes. Soil has been subjected to the action of air and water, which have altered and removed some of the original components so that the proportions of the various substances in the soil are not the same as in the parent rock. The mineral particles constitute the basis or foundation of the soil but not the whole of it.

Each kind of soil has its own individuality. The distinctive feature of this individuality is the soil profile, which consists of a series of layers different from one another in color, texture, structure, and other ways. Each layer is called a horizon, and the succession of layers down to the weathered rock forms the complete soil profile. The profile is best studied in soil that has not been disturbed by cultivation, because horizons near the surface become destroyed by mixing, which occurs during cultivation.

Kinds of Organisms in Soil

Generally speaking, soils are excellent culture media for the growth of many kinds of organisms. This is especially true of the cultivated and improved soils. The microscopic life of soil includes bacteria, yeasts, molds, algae, diatoms, and protozoa. The latter includes amoebae, flagellates, ciliates, and rotifers. In addition, there are present various nematodes, insects, etc.

According to McCalla and Goodding, each spoonful of an arable soil may contain billions of living microscopic organisms. Multiply this by the number of spoonfuls of soil in an acre, and you have figures that are astronomical. The bacteria in an acre of soil of average

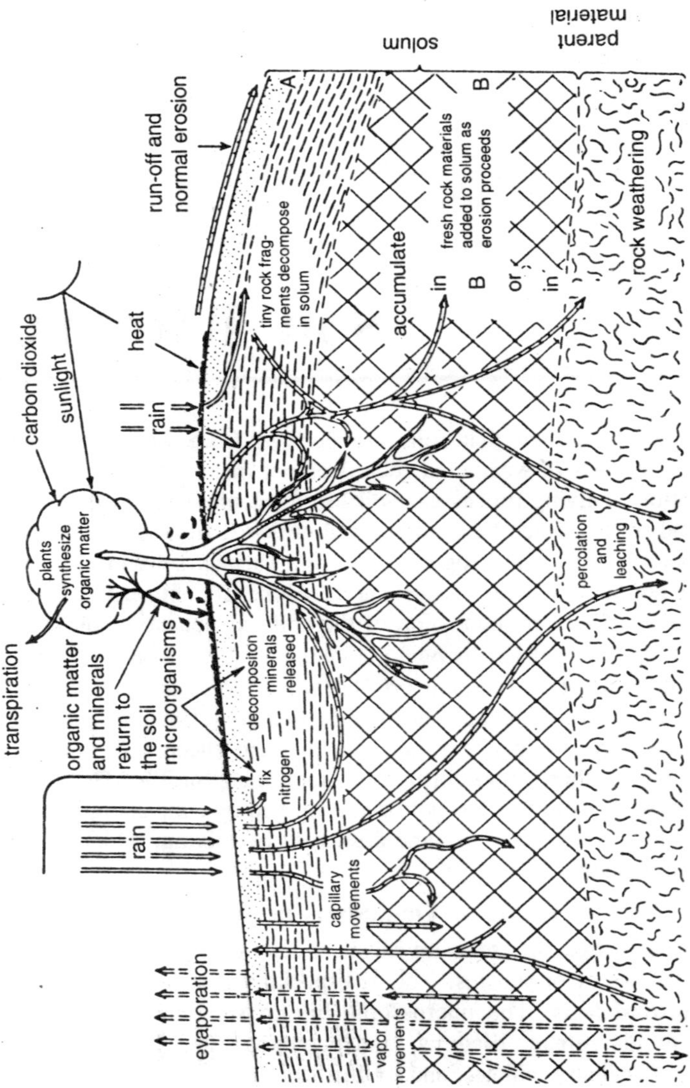

Fig. 5.1. Diagram showing the interlocking cycles leading to the conversion of rock into soil, making for the maintenance of a cover of vegetation on the ground and the maintenance of the soil.

fertility would weigh as much as a medium-sized dairy cow. This seething mass of microorganisms constitutes a crop of 3 to 5 tons per acre-foot of soil that the farmer grows beneath the surface in addition to the crop that he grows above the ground. If this crop of microorganisms beneath the surface is not fed adequately, the crop above ground may suffer from competition, disease, or other adverse effects of the microorganisms. Without the microorganic life, the soil would not be the dynamic, perpetual system that sustains all plants and indirectly all animal life.

Since most of the soil inhabitants are aerobic, the organisms are found in greater numbers in the surface layers. The numbers decrease as the depth of the soil increases. A well-aerated soil contains more organisms than one lacking an abundance of oxygen. The numbers and kinds of organisms found in soil depend upon the nature of the soil, depth, season of the year, state of cultivation, reaction, amount of organic matter, temperature, moisture, etc.

Methods are available for counting the organisms in soil as well as for isolating the various species in pure culture. Since the organisms may vary considerably in their growth requirements, many types of culture media must be employed. The organisms may be aerobic, anaerobic, or facultative types.

Many of the species of organisms present in soil grow in associations with others. The phenomena of symbiosis, synergism, commensalism, and antagonism are believed to be of common occurrence. Two species growing together frequently elaborate metabolic waste products not produced by either organism when grown in pure culture. A product of metabolism of one organism may serve as a nutrient for another species. Antagonistic organisms are usually present and serve to combat other species, especially certain important plant pathogens. This explains why it is exceedingly difficult to determine from laboratory studies on pure cultures what actually takes place in the soil.

In general, the same media and methods employed for the cultivation and isolation of heterotrophic bacteria are used for the propagation and separation of the majority of organisms found in the soil. However, the soil contains some species that are specific and do not grow on the usual culture media. Special media and methods must be employed for their cultivation. These include symbiotic nitrogen-fixing bacteria, nonsymbiotic nitrogen-fixing bacteria, sulfur-oxidizing forms, sulfate-reducing species, urea-decomposing bacteria, cellulose-decomposing organisms, and ammonia-oxidizing species.

Functions of Organisms in Soil

One of the important functions of soil organisms is to decompose various kinds of organic matter of plant and animal origin. This includes stable manures, green manures, plant stubble, plant roots, organic fertilizers, and other products. The decomposition of such compounds is the result of the activities of bacteria, molds, protozoa, worms, and other organisms present in the soil. Each group selects certain constituents of the organic matter suitable for synthesizing its own characteristic protoplasm.

The organic compounds added to soil as a result of biological action include various sugars, amino acids, pentosans, celluloses, lignins, proteins, fats, waxes, tannins, and pigments. These compounds are further decomposed, resulting in the liberation of soluble organic and inorganic constituents. Some of the inorganic compounds, notably ammonia, may be utilized by plant life as a source of nitrogen.

Organic materials, especially stable and green manures, are said to produce four distinct effects upon soil processes and upon plant growth:

1. They supply inorganic nutrients to plants, especially nitrogen and phosphorus.
2. They affect the physical conditions of the soil, especially the moisture-holding and buffering capacities.
3. They supply certain specific elements that may be limiting factors for the growth of some plants.
4. They favor the development of organisms that secrete substances antagonistic to the growth of certain specific forms responsible for plant diseases.

Higher plants are indispensable to human welfare, and the activities of microbes in the soil are intimately related to plant growth. If it were not for the growth of organisms in soil, all animals, including man, and plants would soon perish. Therefore, life could not exist in the complete absence of soil microorganisms.

Biology of Soil

Soil structure has a significant influence on erosion, water intake, and crop growth. A stable granulated soil will permit rapid water uptake, drainage, aeration, and beneficial microbic activity, whereas a dispersed or compact soil has a low infiltration rate. Crops grow more luxuriantly in fertilized, well-aerated, drained, and granulated soils. The most important factors in maintaining or improving the granulation of soils are the presence of microorganisms and their decomposition products.

According to McCalla, "Growing plants and microorganisms constitute the living phase of the soil. In it are living and nonliving bacteria, viruses, bacteriophages, fungi, algae, protozoa, earthworms, nematodes, and insects. These organisms have numerous enzymatic systems which are capable of producing available nutrients for the growing plants by breaking down some of the many organic compounds in the soil.

Physically, a good soil is made up of about 5 per cent organic and 95 per cent inorganic materials by weight. The influence that the small amount of organic matter in most soils exerts is out of proportion to its weight and is more directly related to the volume it occupies. This organic fraction of the soil is made up of plant and animal remains in all stages of decomposition.

Organisms and their metabolic products are active in the formation and stabilization of soil structure, although they play a greater role in stabilization than in formation. A good soil structure is essential for the smooth integration and performance of the multi-groups of organisms in the soil. The beneficial organisms decay plant and animal residues, liberate plant nutrients, and produce soil-structure-stabilizing substances. A deterioration of soil structure results in suppressing the activities of the beneficial groups and promoting the growth of microbial groups that render nutrients unavailable. An unfavorable plant root medium soon develops under anaerobic conditions.

Soil structure affects plant growth in numerous ways. For example, it increases water intake and drainage. It has direct effect on plant growth, such as root penetration, aeration, or anchorage. It influences biological activity such as nitrogen fixation, nitrification, or organic matter decay. A good structure holds to a minimum the processes of anaerobic microorganisms, such as the reduction of sulfates or nitrates to forms unavailable or toxic to the plants.

A desirable soil structure can be maintained by following a rotation including a sod crop, and frequent liberal use of manures and crop residues.

Nitrogen Cycle

Nitrogen is the cornerstone of the structural requirements of all living cells. It is, therefore, absolutely necessary for the growth of bacteria and other organisms. Without an available supply of this element, life could not endure.

A continuous transformation of nitrogen takes place in the soil by various groups of organisms. In the breakdown of protein, the first step is a hydrolysis of the molecules to their constituent building stones,

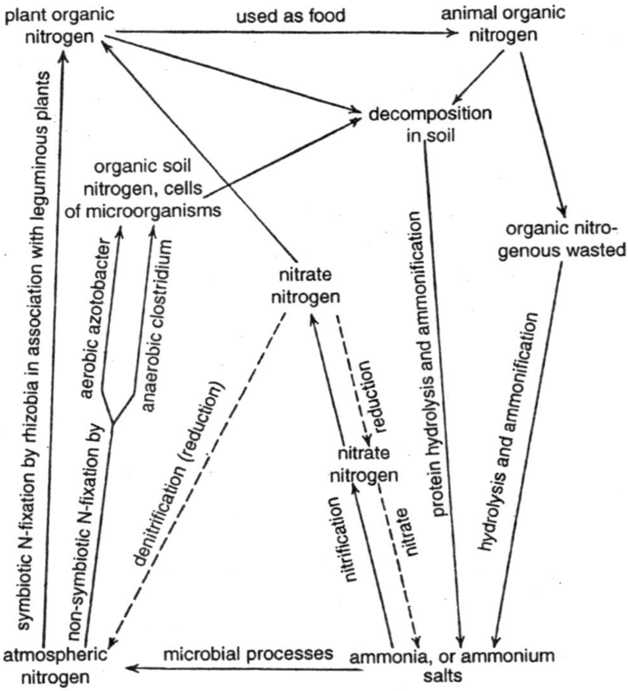

Fig. 5.2. The nitrogen cycle.

or amino acids. The amino acids are then deaminized with the liberation of ammonia:

1. Proteins + H_2O → amino acids.
2. Amino acids + H_2O → ammonia + carbon residue.

After ammonia has been liberated from various nitrogenous compounds, it may be (1) assimilated by soil organisms and again synthesized into proteins, (2) used by higher plant life as a source of nitrogen, (3) absorbed by the colloidal substances in soil and bound as ammonia, (4) acted upon by other soil forms and oxidized first to nitrites and then to nitrates. The organisms responsible for this last set of reactions belong to the autotrophic group of bacteria and are incapable of utilizing organic compounds for structural or energy purposes.

The nitrates may be utilized by various microorganisms and by higher plants and synthesized into proteins, or they may be reduced first to nitrites and finally to free nitrogen. The free nitrogen is lost as far as being available to plant and most microscopic soil life.

However, certain bacteria found in the soil have the ability to utilize free nitrogen and make it available to plant life.

Two groups of organisms are responsible for nitrogen fixation. The organisms in one group are nonsymbiotic nitrogen fixers; those in the other group fix nitrogen only when growing in symbiosis on the roots of certain plants. The free nitrogen is transformed into organic compounds which are made available to plants and soil organisms.

Quantitative Examination of Soil

Soil is the natural habitat of a considerable number of bacterial species, varying from strict autotrophic to heterotrophic forms. Because of this fact, many types of culture media are required to obtain an accurate appraisal of the numbers of organisms present. Such a procedure is out of the question where a rapid determination of the approximate numbers present in a sample of soil is desired.

Soil Colloids. Bacterial organisms are found chiefly in the layer of colloidal material surrounding the inorganic particles of soil. Conn and Conn, McCalla, and others noted that bacteria grew better in the presence of colloidal clay than in its absence. McCalla attributed the stimulation of bacterial growth to the catalytic effect of the clay in speeding up biochemical reactions, either by providing a more efficient utilization of nutritive material or by decreasing the toxic effects of waste products by adsorbing them.

Bacteria in contact with soil colloids adsorb cations. Under normal conditions, the solid material in the soil constitutes about 80 per cent and the water content about 20 per cent. The water is present around and between the particles of soil. Normally, the bacteria probably live in the water films that adhere to the surface of the colloid particles containing adsorbed ions. To quote from McCalla, "The bacterial cell is undoubtedly in close proximity to the soil particles and, assuming that the bacteria may adsorb ions and hold some of them in the outer surface of the cell, this would permit contact exchange of adsorbed ions. Ions with large oscillating volumes would overlap, and exchange between systems could readily take place. Other ions which are strongly adsorbed would not be expected to wander far from the surface of the colloid. On the basis of the displacement of adsorbed methylene blue, the ions would be expected to be adsorbed by the bacteria from the soil colloids in the following series: H > Al > Fe > Mn >Ba > Ca > Mg > K NH, > Na.

From the foregoing facts and theoretical considerations it is suggested that in the adsorption of nutrients from the soil by the

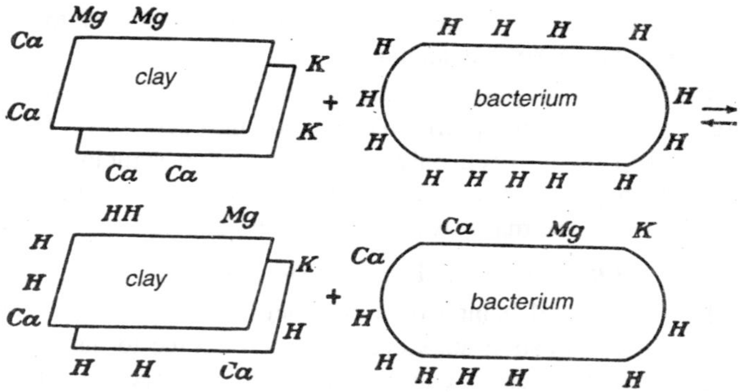

Fig. 5.3. Mechanism of ion exchange between soil bacteria and collidal clay.

bacteria, and possibly by living cells in general, an exchange of adsorbed bases takes place between the bacteria and soil colloid as depicted in Fig. 224. In bacterial metabolism large amounts of carbon dioxide and water are formed. In the presence of H20, H-ions are produced from the carbon dioxide which may be adsorbed at the cell's surface. When a colloidal clay particle, saturated with adsorbed bases, contacts a bacterium saturated with H-ions, an exchange of ions takes place until an equilibrium is reached. As this equilibrium is upset by the more complete utilization of the adsorbed basic ions in bacterial metabolism, a further exchange may take place, the colloidal clay functioning as a constant reservoir for basic ions utilized in the growth of such organisms."

It is difficult to remove or separate the organisms from this colloidal layer. This means that the number of colonies appearing on an agar plate is not an accurate index of the population of a soil sample. Also, no single culture medium is satisfactory for the growth of all species present in soil.

Two methods are generally employed for estimating the numbers of microorganisms in soil: (1) the agar plate method and (2) the direct microscopic method.

Agar Plate Method

The procedure is as follows: A weighed sample of soil is mixed with a known volume of sterile water contained in a screwcap bottle. The sample is vigorously shaken to separate as many organisms as possible from the colloidal material surrounding the soil particles. After the coarse particles have separated, a series of dilutions is prepared from the suspension. Aliquot portions from each dilution are transferred to Petri dishes and mixed with melted agar. The plates are incubated at 25°C. for 2 to 14 days. The colonies are counted and the results expressed as the number of organisms per gram of soil.

Numerous media may be used for soil by the agar plate method. isolating soil bacteria. A medium containing a small amount of organic matter is more satisfactory than one containing an abundance of this constituent. Peptone in media is particularly unsatisfactory as it permits the overgrowth of spore formers and proteolytic *Pseudonwonas,* which prevent the development of the slowgrowing bacteria. The same unsatisfactory result occurs if the temperature of incubation is higher than 25°C.

Objections to the agar plate method

The enumeration of the soil population by this method presents several serious errors. Obligate anaerobes do not grow in the presence of oxygen; autotrophic bacteria fail to multiply in organic media; nonsymbiotic nitrogen fixers grow to a limited extent only; many cellulose-decomposing organisms do not grow on the commonly used media; sulfate reducers grow only in a medium containing sulfate; etc. The counts represent only a fraction of the total bacterial population of soil. The method does give an estimate of the number of organisms in soil capable of growing on a nutrient agar medium.

Molds are commonly present in soil, although the numbers appearing on agar plates represent only a small percentage of the total counts. Yeasts are less prevalent than molds, increasing in numbers in acid soils and in soils of orchards and vineyards.

The counts by the agar plate method vary from 200,000 to 100,000,000 per gram of soil. The colonies appearing on plates consist of about 10 to 40 per cent actinomycetes, 50 to 80 per cent non-spore-formers, and 3 to 10 per cent spore bearers.

Direct Microscopic Method

The method is as follows: One part of soil is suspended in 10 times its weight of a 0.015 per cent aqueous solution of agar. The purpose of the agar is to fix the organisms to the slide. An area of 4 sq. cm. is ruled off on a microscope slide by means of a china-marking pencil. One-tenth milliliter of the soil suspension is transferred to the ruled area and spread out uniformly. The slide is dried on a flat surface over a water bath, and then covered with a solution consisting of 1 per cent rose bengal dissolved in a 5 per cent aqueous solution of phenol. In a good preparation, the bacteria take a deep pink or red color, whereas the mineral constituents do not stain. Some of the dead organic matter appears light pink, but most of it stains either yellow or not at all. If the bacteria appear faintly stained or if everything is colored pink, a new preparation should be made. The former condition generally means that the slide has been washed too long; the latter indicates that the staining solution is too old.

The slide is examined under a calibrated oil-immersion objective and the number of organisms per field counted. At least 25 fields are counted and an average taken. Knowing the area and count per field, it is a simple matter to calculate the number of organisms present in a gram of soil.

The direct counts are from 5 to 20 times greater than by the agar plate method. Much of the discrepancy is due to bacteria which fail to grow on the plates rather than to clumps of organisms which do not break up in the plating method.

Objections to the direct method

The method records dead organisms which do not develop on agar plates. The organisms must be evenly distributed over the slide. It is advisable to prepare several slides and to take an average of the counts. A disturbing factor is the difficulty of recognizing the bacteria. Many soil forms are too small to be easily distinguished from soil particles. Also many soil particles resemble bacteria. Therefore, considerable experience is necessary in making accurate determinations of the soil population by the direct method.

Variations in Soil Counts

The numbers of organisms in soil are not uniform even over a very small area. In order to increase the accuracy of the determination, several samples from the same plot should be collected and an average taken of all determinations for computing the final count. A single determination may be considered as valueless for calculating the soil population.

It has been reported that if soil is partially sterilized with steam, or with a volatile antiseptic such as toluene, the bacterial count first decreases, followed by a sharp increase in numbers and activity. If a little untreated soil is added, the bacterial count again decreases. The partial sterilization of the soil destroys the protozoa but not the bacteria and is followed by a sharp increase in the latter, The addition of an untreated soil reintroduces protozoa and results again in a decrease in bacterial numbers. The important protozoa responsible for the daily variations in bacterial numbers include ciliates, flagellates, and rhizopods. These organisms are widely distributed in soils.

It has been shown that an inverse relationship exists between the numbers of bacteria and active amoebae in soil. A decrease in numbers of protozoa is followed by an increase in the bacterial population and vice versa. This means that if a soil determination is to be of any value, it is *very* important that samples be examined from different areas of the same plot as well as at frequent intervals over a long period,

Soil bacteria differ in their edibility by protozoa. Certain groups, such as the root-nodule bacteria, generally are resistant to attack by micropredators, whereas some strains of *Aerobacter* are attacked by all soil protozoa which have been studied,

Qualitative Studies of Soil Organisms

Lochhead showed that the qualitative nature of the soil microflora was markedly influenced by the growing plant. In the rhizosphere (the zone influenced by root excretions), the Gram-negative rods were increased in numbers, whereas the Gram-positive rods, cocci, and spore bearers were less abundant.

Most rhizosphere bacteria are saprophytes. Some live on the root surface; others penetrate the roots. Some are restricted to the cortical cells, whereas others go deeper, passing between the cells and invading them. Some are harmless; others are destructive or have a favorable effect on the development of the host.

Bacteria are generally more numerous in the rhizosphere than in soil more distantly removed. They differ from those beyond this zone in being physiologically more active. Also they show (1) a greater proportion of motile forms, (2) a pronounced increase in chromogenic bacteria, and (3) a higher incidence of gelatin liquefiers and glucose fermenters.

This increased growth is unquestionably related to the supply of inorganic and organic nutrients at the root-soil interface. Plant excretion of inorganic substances, growth factors, amino nitrogen, and a variety of other substances have been generally accepted. Also the food materials supplied by sloughed-off root caps, root hairs, and cortical and epidermal cells should not be overlooked, since they are rapidly attacked by soil organisms and supply considerable quantities of organic and inorganic nutrients.

Plant roots affect microbial growth, and the plant in turn is affected by the increased activity of the microbial population of the rhizosphere. The information available is inadequate to indicate whether plant development is increased or impaired by the presence of the rhizosphere organisms.

Classification of Soil Bacteria

Winogradsky placed soil bacteria into two groups which he named the zymogenous and the autochthonous types.

Zymogenous Bacteria

The zymogenous types consist of actively fermenting organisms requiring for their activity ingredients which are quickly exhausted. They are involved in processes in which organic matter is made available to plants. The added organic matter is rapidly attacked in successive stages, each of which involves a specific group of organisms. The organisms increase rapidly whenever furnished with the special nutrients to which they are adapted, then return again to low numbers until another occasion for active growth occurs. Some of these organisms require special media for laboratory cultivation; others grow on ordinary media. Organisms of this group include the nitrifiers, nitrogen fixers, cellulosehydrolyzing bacteria, sulfur-oxidizing forms, acid-fast bacteria, molds, streptomycetes, spore-forming *Bacillus* species, and non-spore-forming *Pseudomonas or* closely related species.

Autochthonous Bacteria

The autochthonous types are more fami-liarly known as the indigenous or native bacteria. They are found in soil in fairly high and quite constant numbers. They do not change in numbers by the presence or absence of specific nutrients, The exact function of these organisms in soil is not clearly understood. They appear to utilize the same nutrients (salts) as higher plants and display a low level of activity in soil. They utilize the same sources of nitrogen as plants and may be looked upon as rivals. Since bacteria are short-lived and are readily autolyzed after death, their rivalry with plants cannot be considered serious, The autochthonous bacteria produce weak fermentation reactions which make their classification difficult. Because of this fact, they are classified almost entirely on morphology.

The group includes some of the streptomycetes, which are more appropriately classified here rather than under the zymogenous bacteria.

A large number of closely related soil species have characteristics intermediate between the true bacteria and the streptomycetes. The organisms are somewhat pleomorphic. In young cells they appear as rods which may vary in size and shape from straight .to bent, curved, swollen, or club-shaped forms. Short filament formation with rudimentary budding may occur. These organisms are placed in the genus *Arthrobacter.* They resemble streptomycetes only in their ability to produce occasional short mycelia.

A number of small, nonpleomorphic, nonsporing, motile rods are found in soil; they differ from the *Arthrobacter* in showing no tendency toward filament formation or the appearance of coccoid forms. These organisms are placed in the genus *Agrobacterium.* Most of these species are plant pathogens.

Production of Ammonia

Most species of bacteria are capable of decomposing proteins and protein-split products with the liberation of ammonia as one of the compounds. The ammonia is released prior to the utilization of the carbon residue chiefly for energy, The production of ammonia is an essential stage in the formation of nitrate in soil. Most plant crops are largely dependent on soil nitrates for structure and growth.

The decomposition of nitrogenous compounds for energy takes place only in the absence of a rapidly utilizable carbohydrate. In the presence of a rapidly fermentable carbohydrate, bacteria derive their energy from this source, utilizing the nitrogenous compounds for structure only. This results in a greatly lowered production of ammonia and nitrate. Therefore, maximum ammonia production takes place only in the absence of a rapidly fermentable compound.

Urea-hydrolyzing Organisms. Other compounds besides proteins and their degradation products are capable of yielding ammonia by bacterial action. These include nitrogenous compounds of a nonprotein nature.

Stable and barnyard manures are often used as fertilizers because of their nitrogen content. They help to replenish the nitrogen supply of the soil. A high content of urine is often present in such waste material. The most important nitrogen compound present in urine is urea. Many organisms have the power of converting the urea to ammonium carbonate and finally to free ammonia and carbon dioxide, according to the equation

$$O{=}C\begin{matrix} \diagup NH_2 \\ \diagdown NH_2 \end{matrix} + 2H_2O \rightarrow (NH_4)_2CO_3$$

$$(NH_4)_2CO_3 \rightarrow 2NH_3 + CO_2 + H_2O$$

The presence of urea-hydrolyzing organisms in soil is determined by incorporating urea in a culture medium. The liberation of ammonia from the medium indicates the presence of one or more organisms capable of attacking the compound. The important urea-decomposing organisms include *Bacillus pasteurii,* a Gram-variable spore-forming rod with rounded ends, usually not in chains, producing round, terminal to subterminal spores which cause a bulging of the sporangia, optimum temperature 28 to 35°C; *blicrococcus ureae,* a Gram-variable, nonmotile, aerobic coccus, appearing singly, in pairs and clumps, optimum temperature 25°C; *Sarcina ureae,* a Gram-positive coccus, occurring singly, in pairs, and in packets, motile by means of a single flagellum, endospores of an unusual type, located centrally, resist heating to 80°C for 10 min., optimum temperature 20°C.

The urea bacteria thrive best in media containing urea, especially when made alkaline with ammonium carbonate. The organisms are capable of converting urea rapidly to ammonium carbonate. They are commonly found in air, water, soil, and manure. About 2 per cent of the organisms present in surface soil and about 10 per cent in manure are capable of decomposing urea.

In addition to the above, some members of the genus *Proteus may* utilize urea. This applies also to many coliform and paracolon cultures, especially to the *Aerobacter.*

Digestion of Cellulose

The greater part of the organic matter in soil is decomposed by bacteria in the process of acquiring energy. The simpler carbohydrates and some of the polysaccharides are attacked by a large number of soil bacteria. The addition of such compounds to soil causes a rapid increase in the bacterial population.

Cellulose constitutes the chief part of the solid framework (cell walls) of plants. It is one of the most important constituents added to soil. Under normal conditions of temperature and moisture, cellulose disappears almost completely and quite rapidly.

The cellulose molecule is built up of units of β-glucose. Two molecules of β-glucose are combined through a 1,4-linkage to give 8-cellobiose. The cellulose molecule consists of a chain of cellobiose units linked end to end through 1,4-,3-glucosidic linkages:

The first stage in the utilization of cellulose is a hydrolysis of the molecule to cellobiose, and secondly to glucose. Two enzymes, cellulase and cellobiase, are concerned in the reaction. Cellulase hydrolyzes cellulose to cellobiose, and cellobiase splits cellobiose to two molecules of glucose. The two enzymes cellulase and cellobiase accompany each other. They are of tremendous importance in the dissolution of insoluble cellulose in the soil. Since animals do not elaborate a cellulase, the

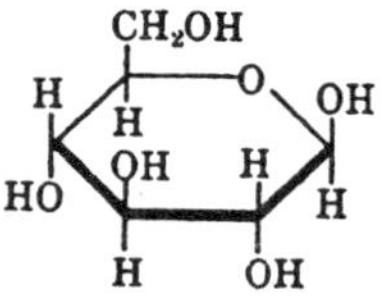

β-Glucose

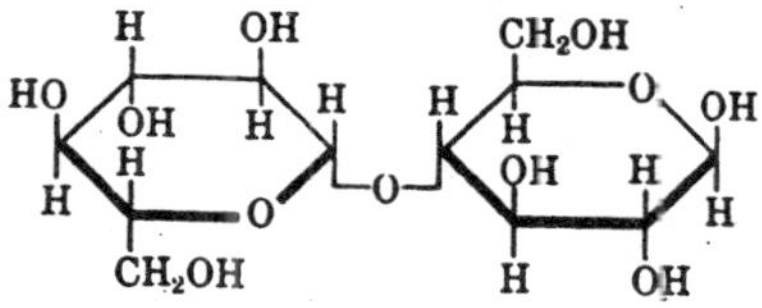

β-Cellobiose

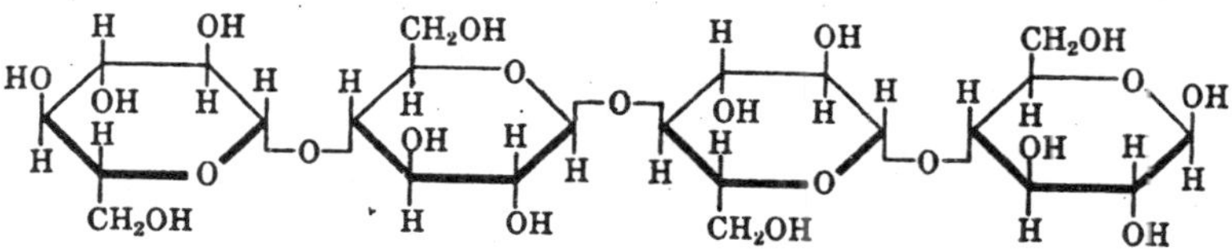

Cellulose chain

presence of cellulose-decomposing bacteria in the intestines of herbivorous animals is responsible for the hydrolysis of some cellulose.

Organisms Involved. Cellulose is attacked by a wide variety of soil bacteria, including (1) aerobic mesophilic species, (2) aerobic thermophilic forms, (3) anaerobic mesophils, and (4) anaerobic thermophilic types. The organisms utilize cellulose as a source of energy. In addition to the bacteria, a number of molds and protozoa are able to attack cellulose.

Aerobic Species

Almost all of the aerobic mesophilic cellulose bacteria so far isolated have been placed in the genera *Cell falcicula* (short rods or spindles with pointed ends, Gram-negative); *Cellulomonas* (small, pleomorphic rods, Gram-negative); *Cellvibrio* (long, slightly curved rods with round ends, Gram-negative); *Cytophaga* (flexible, sometimes pointed rods, Gram-negative); *Pseudomonas* (rods, sometimes coccoid, Gram-negative); *Sporocytophaga* (vegetative cells rod-shaped, microcysts spherical); *V ibrio* (short, curved rods, Gram-negative). With the exception of *P. erythra,* all are capable of growing well in the presence of a variety of carbohydrates.

The presence of aerobic organisms can be easily demonstrated by adding soil to a culture medium composed of certain inorganic salts and cellulose or other satisfactory carbohydrate (filter paper, cotton, starch) and incubating for several days at the proper temperature. The carbohydrate is slowly attacked and liquefied.

Anaerobic Species

The anaerobic cellulolytic bacteria may be placed in the following groups: mesophilic non-spore-forming rods and cocci, mesophilic spore formers, thermophilic spore producers, and streptomycetes. Most of the cellulolytic bacteria encountered in nature are anaerobic spore producers of the genus *Clostridium.* Some anaerobic species are capable of dissolving -cellulose at temperatures of 60 to 65°C. These organisms are widely distributed in nature, being found in soil, river mud, and the intestinal contents of animals. They may be isolated by inoculating an inorganic medium containing cellulose with an infusion of rapidly decomposing manure and incubating anaerobically at 65°C. If the organisms are present, the cellulose will be dissolved in 6 to 8 days.

Hungate isolated a cellulolytic streptomycete from the alimentary tract of termites. The organism produced chiefly propionic acid with smaller quantities of acetic acid and carbon dioxide. The organism proved to be a species of *Micromonospora.*

The end products of bacterial action on cellulose consist chiefly of acetic, butyric, formic, lactic, propionic, and succinic acids; alcohol; carbon dioxide; and hydrogen. Cellulolytic organisms show considerable variation in the kinds and quantities of end products produced.

Other Organisms

A considerable number of molds and mold-like organisms have been described as being capable of utilizing cellulose. The species that have been most extensively investigated are members of genera *Aspergillus, Penicillium, Monilia, Alternaria, Trichoderma, Mucor, Rhizopus, Merulius, Paxillus, Fusarium, Cladosporium, Chaetomium, Polyporus, Actinomyces,* and others. Molds are commonly found growing on old paper, especially if kept in a damp or humid atmosphere. They have been responsible for the destruction of many valuble manuscripts.

Specificity of the Reaction

It is generally believed that cellulolytic organisms must have cellulose for energy and cannot substitute any other carbon source. This statement is not strictly correct. Glucose and other carbohydrates have replaced cellulose in the cultivation of a number of species of *Cytophaga.* It is highly probable that glucose can replace cellulose in the metabolism of other cellulose decomposers. The similarity in rates of cellulose, cellobiose, and glucose oxidation *by Cytophaga hutchinsonii* provides good evidence that cellulose is first hydrolyzed to glucose, followed by the utilization of the sugar for energy.

Symbiotic Nitrogen Fixation

It was shown several generations ago that the growth of certain plants in soil resulted in the stimulation of the succeeding plant crop.

The fertility of the soil was increased. The plants responsible for this stimulation were almost entirely members of the family Leguminosae.

Hellriegal and Wilfarth demonstrated that the stimulation was due to an increase in the nitrogen supply of the soil, which resulted from the presence of small tumor-like growths or nodules on the roots of the plants. In the absence of nodules, no stimulation of growth of the succeeding plant crop occurred.

The formation of nodules on roots is the result of the associated growth of plant and bacteria. The bacteria are members of the genus *Rhizobium* and are commonly referred to as the root-nodule bacteria. The organisms live in the cells of the plant roots, where their growth and metabolic activities cause a swelling or nodule to form on the root. The organisms utilize the free nitrogen of the atmosphere and synthesize it into nitrogen compounds. The plant obtains its nitrogen from the synthetic activities of the bacteria, while the organisms derive their food from the plant. The plant and bacteria live together for mutual benefit. Such an association is known as symbiosis.

Formation of the Nodule

The nitrogen-fixing organisms live in the soil in the free state. Recent work indicates that they fix nitrogen only when growing in association with the plant. They gain entrance to the plant through the root hairs or other epidermal cells. The bacteria multiply very rapidly, forming long filaments in the root hairs and into the parenchyma of the root. The organisms cause a rapid proliferation of the surrounding tissue in the innermost cells of the root cortex, which results in the formation of a young nodule. The young nodule pushes out the overlying parenchyma and epidermis and produces a swelling on the side of the root. A nodule consists of a mass of thin-walled parenchyma cells, which are usually almost filled with the specific organism. A corky layer and branches of a vascular system are also present. This system provides the bacteria with their nutrients, and the plant in turn takes away the nitrogen compounds synthesized by the bacteria.

Nodules vary in shape and size according to the plants on which they occur. Nodules on clover are round or oval; those on pea are commonly round, elongated, and frequently clustered; on the bean and soybean plants nodules are relatively large, round, and firmly attached to the root; those on alfalfa are usually long, finger-like growths.

Nitrogen Fixation in Excised Nodules

Until comparatively recently, nodules removed from roots and placed in culture solutions were reported as being unable to fix nitrogen. Aprison and Burris placed excised nodules under an atmosphere containing N^{15}. After exposure for 30 min., the nodules were ground in the presence of acid, the pulp centrifugated, and the supernatant liquid analyzed for N^{15}. Fixation was readily and consistently demonstrated.

Plants Involved. Nodule formation occurs most successfully on the members of the family Leguminosae. However, there are some species in the family which fail to develop nodules. Moreover, a few nonleguminous plants show the presence of nodules.

According to Allen and Allen, the family Leguminosae contains 429 genera comprising approximately 10,000 species of plants. Of these, 179 genera (41 per cent), containing 946 species (9.4 per cent), have received attention regarding nodulation. Only 77 leguminous species have been consistently listed as lacking nodules.

Organisms

Great masses of organisms are present in the nodule. Their presence may be easily demonstrated by crushing a washed nodule between two glass slides, fixing the smear, and staining by the usual technique. It is a relatively simple matter to isolate a pure culture of the organism from a previously washed and sterilized nodule. The usual culture media are not satisfactory for the cultivation of the organisms. A medium that has yielded very good results is known as Ashby's mannitol phosphate agar, which is an inorganic medium to which is added mannitol as a source of energy. Colonies appear in 5 to 10 days when the plates are incubated at 25°C.

The members of the genus *Rhizobium* are rod-shaped, measuring 0.5 to 0.9 by 1.2 to 3 μ, being motile when young. Cells from nodules are commonly irregular with X-, Y-, star-, pear-, and club-shaped forms. Swollen or vacuolated forms sometimes predominate. These forms are also known as bacteroids. Round or coccus cells may also be present. Cocci arise from the swollen or vacuolated cells, which later return to the rod form. Some have described the various forms as orderly stages in the life cycle of the organism. It is Gram-negative and aerobic. The genus includes six species: *Rhizobium leguminosarum, R. phaseoli, R. tri f olii, R. lupini, R. japonicnrn,* and *R. meliloti*. A specific name has not been given to the organism that produces nodules on plants of the so-called "cowpea" group. Walker and Brown proposed a consolidation of the soybean and cowpea groups as being inoculated by a single species, *R. japonicum*. Reid and Baldwin would include the lupine group also. The colonial characteristics of the organisms show some variation, depending upon the plants from which they are isolated. Some species show raised, glistening, semitranslucent, white, slimy, and occasionally viscous colonies with considerable gum formation. In other species the colonies appear small, slightly raised, glistening, opaque, white, and butyrous with little gum formation.

Although the bacteroid form predominates in the nodule, it is usually absent when nodular material is streaked on agar plates. After death and dissolution of the nodule, the organisms pass back into the soil as rod-shaped or ovoid cells which can develop flagella. In this stage the organism is able to swim in the soil moisture, and it is probable by such means that it reaches the roots of its host plant.

Cross-inoculation Groups

There are almost as many strains of rootnodule bacteria as susceptible plants. Nitrogen fixers isolated from the nodules of one plant are not necessarily capable of producing nodules on all legumes. The clover plant grows best when inoculated with nodule bacteria from clover; peas grow best with bacteria from peas; etc. The right pair must be brought together for successful results. The plant and rhizobia must be highly compatible. The fact that legume bacteria are so highly selective was responsible for the recognition of so-called crossinoculation groups of leguminous plants. Any plant within such a group is inoculable with a culture of the proper organism, usually comprising several strains known to inoculate effectively all the legumes in that particular group.

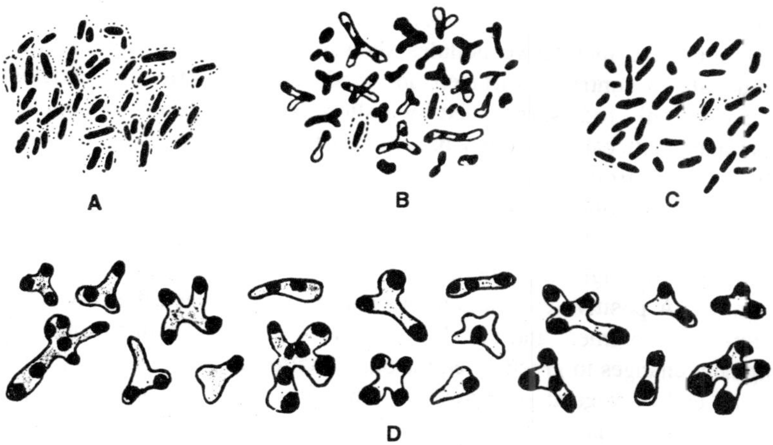

Fig. 5.4. Rods and bacteriods from root nodules. A, rods from a white ineffective nodule; B, bacteroids from a red effective nodule; C, rods from a green ineffective nodule; D, various forms of bacteroids.

Cross-inoculation groups, however, have been open to criticism because some strains of bacteria are effective for more than one group. But for all practical purposes these groups offer a convenient and workable plan for the preparation of inoculant cultures. Farmers have been accustomed to ordering legume cultures according to group designations. Seven groups are generally recognized: alfalfa, bean, clover, cowpea, lupine, pea and vetch, and soybean. Wilson distinguished 22 cross-inoculation groups among the legume bacteria. The causes of this host-plant specificity are unknown.

Organisms in the various cross-inoculation groups show some differences in their fermentation reactions. At best, however, attack on carbohydrates is feeble. With some strains this is probably due to their slow rate of growth on artificial media.

Some of the root-nodule bacteria show a strong tendency to produce large quantities of gum when cultivated on carbohydrate media. Gum formation may be so pronounced as- to render the medium strongly viscid. Organisms in different cross-inoculation groups vary considerably in their capacity to produce gum. The gum is nitrogen-free, water-soluble, and precipitated from solution by alcohol or acetone. On heating with a mineral acid, the gum is hydrolyzed to a reducing sugar.

Mechanism of Symbiotic Nitrogen Fixation

Virtanen observed two types of nodules on pea plants. One type effectively fixed nitrogen, whereas the other did not. The plants containing the ineffective nodules showed no better growth than the uninoculated controls. Virtanen observed that the ineffective nodules contained only rods surrounded by a slimy layer; the effective nodules contained noticeably swollen bacteroidal forms. He believed that the slimy layer around the ineffective organisms interfered with their nutrition and oxygen uptake.

The root nodules of leguminous plants that actively fix nitrogen contain a red pigment, whereas the ineffective nodules are devoid of pigment. A positive correlation appears to exist between the concentration of red pigment and the ability of the nodules to fix nitrogen. The pigment changes to green when nitrogen fixation ceases in annual plants at the end of vegetative growth or when the plants are placed in the dark for a few days. The red pigment may be concentrated by treating a suspension of crushed nodules with ammonium sulfate. The pigment is capable of storing and transporting oxygen. Chemically it is a hemoprotein having absorption bands similar to hemoglobin. Virtanen named the pigment leghemoglobin. The green pigment may also be isolated from the green nodules in the same manner as leghemoglobin from the red nodules. It is also a chromoprotein. However, the absorption bands of the green and red pigments are not the same.

A number of theories have been advanced to explain the mechanism of biological nitrogen fixation. The two most important theories appear to be those advanced by Virtanen and Burris and Wilson.

Theory of Virtanen

Legumes were inoculated with an effective rootnodule organism and grown in a sterile, nitrogen-free medium. Nitrogen compounds appeared immediately after nodule formation. The nitrogen compounds were excreted into the soil from the root nodules rather than from the roots. In some cases considerably over 50 per cent of the total nitrogen fixed was excreted, and it was most powerful in young plants when the nodules were still growing and showed no signs of impairment. The excreted nitrogen was chiefly amino-N as L-aspartic and L-glutamic acids. Some 8-alanine was also present which was produced from the L-aspartic acid by decarboxylation.

Theory of Burris and Wilson

Hydroxylamine is taken as the central nitrogen compound from which ammonia is formed through reduction, and which can also react directly with keto acid.

Both theories favor the formation of amino dicarboxylic acids as the primary compounds in biological nitrogen fixation.

Artificial Inoculation of Plants

The efficiency of symbiotic nitrogen fixation has been greatly increased by (1) choice of the proper legumes, (2) development of new varieties of leguminous plants, (3) artificial inoculation of seeds with pure cultures of root-nodule bacteria, and (4) adjustment of the environment to optimum conditions.

Cultures of the rhizobia are frequently added to soil to increase the nitrogen supply. This is generally practiced by soaking the seeds of a legume in a culture of the appropriate species. This procedure has become a well-established practice, especially when a legume plant is seeded for the first time.

For successful results it is extremely important that the strain of organism selected be one that is known to produce effective nodules on the legume. There appears to be no rule to follow in the selection except the actual testing of various strains on the plant to be affected. Greenhouse tests should be supplemented with actual field tests to make sure of the proper selection for the mass production o^c commercial cultures.

The addition of an abundant supply of nitrate to soil results in a luxuriant plant growth. Under such conditions nodules are probably not formed even though the seeds have been soaked in a culture of the organisms. The suppression of nodule formation is due to a change in the metabolism of the plants. On the other hand, the addition of calcium and phosphorus to soil results in a stimulation of nodule production. Other factors influencing nodule formation include reaction, temperature, moisture, and oxygen content of soil.

Nonsymbiotic Nitrogen Fixation

In addition to the organisms discussed in the preceding section, the soil harbors certain free-living bacteria that are also capable of fixing free nitrogen. The first nonsymbiotic nitrogen fixer isolated was *Clostridium pasteurianum,* a motile, Gram-positive, spore-forming, anaerobic rod. It is capable of fixing small amounts of nitrogen. At least 15 other species of clostridia are capable of producing the same reaction. In addition, a number of photosynthetic bacteria of the genera *Rhodospirillum, Chromatium, Rhodomicrobium, Rhodopseudomonas,* and *Chlorobium* have been shown to fix nitrogen gas. However, the most important nonsymbiotic nitrogenfixing organisms are members of the genus *Azotobacter.*

The genus is characterized by relatively large rods or even cacti, sometimes almost yeast-like in appearance. The cells are without endospores. Flagellation is typically peritrichous. The organisms are Gram-negative, obligate aerobes, usually growing in a film on the surface of the medium. They are capable of fixing atmospheric nitrogen when provided with carbohydrate or other energy source. They grow best on media deficient in nitrógen. Soil and water bacteria are included in the genus. Three species are recognized: *Azotobacter chroococcum, A. agilis,* and *A. indicus.*

In the presence of nitrogen compounds such as ammonium salts, nitrate, urea, aspartic acid, and asparagine, the organisms fail to fix nitrogen. However, the relative availability and probable occurrence of fixed nitrogen compounds in soils are such that they would be unlikely markedly to retard nitrogen fixation by *Azotobacter* under most conditions.

The presence of nitrogen-fixing organisms may be demonstrated by adding a small amount of fertile soil to a mineral medium containing mannitol. The medium is incubated at a temperature of about 25°C.

After several days at this temperature, a pellicle begins to form on the surface of the medium. A stained smear of the pellicle material reveals the presence of many typical cells of *Azotobacter*. Depending upon the species, the cells may or may not be surrounded by slime. New cultures are prepared by transferring some of the pellicle to a flask of fresh medium. After several transfers, the culture becomes heavily enriched in *Azotobacter*. A loopful of the pellicle is now streaked over the surface of the same medium solidified with agar. The plate is incubated at 25°C. until colonies appear. A pure culture is obtained by transferring a typical colony to fresh medium.

The organisms are widely distributed in soil. Their absence from certain soils is probably due to an unfavorable pH. Cells of *Azotobacter* cannot develop if the reaction of soil is more acid than pH 6.

The size and shape of cells are subject to great variation, being dependent upon a number of environmental factors, such as composition of the medium, presence of essential minerals, presence of oxygen, temperature of incubation, and reaction of the medium. Esposito and Wilson reported that molybdenum, iron, and calcium were specifically required by one species for the fixation of nitrogen. Ammonium ion eliminated the need for molybdenum and calcium and had a sparing action on the iron requirement.

Mechanism of Nonsymbiotic Nitrogen Fixation

Virtanen believes that the reaction scheme for nitrogen fixation in *Azotobacter* is the same as in *Rhizobium*.

Rhizobium and *Azotobacter* show marked differences in the amounts of their excretion products. Where *Rhizobium* excretes 50 to 80 per cent of the total fixed nitrogen from their nodules, *Azotobacter* excrete only 10 to 25 per cent. Most of the nitrogen fixed by *Azotobacter* is present as cell substance. However, it makes little difference in the end, since the organisms soon die and the total nitrogen content is returned to the soil.

Azotobacter Inoculation of Soil

Many attempts have been made to increase the nitrogen content of soil by inoculation with pure cultures of *Azotobacter*. In most cases no effect on the yield of nonleguminous plants was obtained, even though unusually high numbers of *Azotobacter* were established in soil. In general, the population steadily decreased in the two or three months following establishment of growth. In a few experiments a detrimental influence was noted. Failures have been attributed to (1) absence of suitable environment, such as proper temperature, moisture, oxygen concentration, food, and reaction of soil; (2) absence of a carbon source; and (3) injurious *effects* of end products liberated in the decomposition of added carbohydrate. For more information see Jensen, Wilson and Burris.

Autotrophic Bacteria

Chlorophyllous plants are capable of effecting a synthesis of their own organic compounds. They take carbon dioxide and combine it with water in the presence of sunlight to form carbohydrate. Nitrogen is absorbed from the soil in the form of nitrate or other inorganic nitrogen compound and synthesized into protein. The other necessary elements are also taken from the soil in the form of inorganic compounds. Some bacteria cannot synthesize their complex protoplasm from simple inorganic materials but must have organic compounds such as proteins, peptones, amino acids, and vitamins for growth. These are classified as strictly heterotrophic organisms.

Most bacteria are able to utilize both inorganic and organic compounds for growth. These are facultatively heterotrophic forms and constitute the majority of bacteria found in nature. In addition to the above, some organisms are able to manufacture all new organic cell material from carbon dioxide, as the main source of carbon, and simple inorganic salts. These *are the* strictly autotrophic bacteria.

The number of strictly autotrophic species is small. A greater number of species are facultatively autotrophic, being capable of existing both autotrophically and heterotrophically. The strict forms include the nitrifying bacteria and some of the sulfur and iron organisms. The facultative forms can obtain their energy from (1) the oxidation of inorganic substances and carbon from carbon dioxide or (2) purely organic compounds. The hydrogen bacteria, and some of the sulfur and iron forms, are facultatively autotrophic.

Nitrifying Bacteria

The nitrifying organisms derive their energy from the oxidation of ammonia to nitrite and nitrite to nitrate. They depend on this oxidation for growth. The nitrifiers fail to grow unless the specific nitrogen compounds are available for energy. Also, they must have carbon dioxide as the only source of carbon.

Nitrosification

The oxidation of ammonia to nitrite is called nitrosification. The organisms involved are members of the genera *Nitrosomonas, Nitrosococcus, Nitrosospira, Nitrosocystis,* and *Nitrosogloea.* The cells are rod-shaped, ellipsoidal, spherical, or spirillar; aerobic; nonsporulating.

They grow best at about 25°C. Organisms secure their energy from the oxidation of ammonia to nitrite:

$$2NH_s + 3O_2 \rightarrow 2HNO_2 + 2H_2O + 79 \text{ Cal.}$$

The organisms fail to grow on media containing organic matter unless compounds of ammonia are present.

Isolation of organisms

Nitrite formation takes place best in neutral or slightly alkaline media. Since the organisms are strongly aerobic, cultures should be exposed in shallow layers, preferably in Erlenmeyer flasks. The ammonia is oxidized to nitrous acid. This continues until all the ammonia has been oxidized to nitrite. The addition of more ammonium salt results in the continuation of the process, provided there is an excess of insoluble magnesium carbonate to neutralize the nitrous acid as it is formed. When all carbonate has been decomposed, the organisms should be transferred to fresh medium. Colonies may be obtained by streaking the organisms over the surface of the same medium solidified with silicic acid. Inorganic silicic acid is superior to agar as a solidifying agent for the cultivation of nitrifying bacteria.

The newer methods for the preparation of silica gel plates make use of initially nearly pure silica preparations purified by passage through ion exchange resins followed by sterilization. Since the methods of preparation are lengthy, the reader is referred to the reports of Kingsbury and Barghoorn, Pramer, and Smith.

Nitrification

The oxidation of nitrite to nitrate is known as nitrifi-cation. Some autotrophic organisms cannot oxidize ammonia to nitrite but can oxidize the latter to nitrate.

$$HNO_2 + \frac{1}{2}O_2 \rightarrow HNO_2 + 21.6 \text{ Cal.}$$

These organisms are placed in the genera *Nitrobacter* and *Nitrocystis*. Since nitrite is seldom found in soils, it may be assumed that the two groups of nitrifying organisms nearly always occur together.

The nitrifiers are widely distributed in nature, being found in practically all neutral and alkaline soils. They are the agents primarily responsible for the appearance in soil of nitrate, which is in turn utilized by higher plants in the synthesis of cell material.

Isolation of organisms

The medium used is similar to that employed in the preceding section except that sodium nitrite is substituted for the ammonium salt. Since the organisms are also strongly aerobic, the medium is best dispensed in shallow layers in Erlenmeyer flasks. The medium is inoculated with the soil sample and incubated at about 25°C. Several transfers to fresh medium are necessary to obtain a culture rich in nitrifying organisms. The nitrifying bacteria are peculiarly sensitive to alkalinity, oxidizing little, if any, nitrite to nitrate at values just above pH 7.7. Just below this value, the oxidation of nitrite to nitrate proceeds so rapidly that the purely biological nature of the transformation is open to question.

Sulfur Bacteria

Sulfur is widely and abundantly distributed in nature, being present in the free and combined form. In the combined form it exists in both inorganic and organic compounds. Gases emitted from volcanoes contain sulfur dioxide and hydrogen sulfide. Sulfur is present in large quantities in sulfur springs. Volcanic sulfur usually occurs as a sublimate on the walls of vents, probably as the result of action between hydrogen sulfide and sulfur dioxide. The commercial source of sulfur is chiefly as crude brimstone obtained from the sides of volcanoes or mined in certain parts of the world. Extensive deposits are found on the island of Sicily and in Louisiana. Commercially it occurs as brimstone, or as flowers of sulfur, which is prepared from the crude brimstone. Sulfur finds its way into the soil from the decomposition of native rock, from organic manures, and from rain water.

Sulfur is one of the elements absolutely essential to. living organisms. It enters into the compositions of all plants and animal cells. The addition of sulfur to a soil low in the element results in a marked stimulation of growth. As the plants and animals die and decompose, the sulfur finds its way into the atmosphere. The gases are dissolved by rain water and again returned to the soil. Some soil organisms are able to convert the sulfur-containing gases to sulfates. Some of the sulfates are utilized by growing plants, and some are leached out by waters and carried off to the ocean. The sulfates in the ocean may be reduced to sulfides and then precipitated by iron as iron sulfide, or they may be converted into insoluble calcium sulfate (gypsum), Deposits of gypsum are believed to have been formed in this manner. Deposits of sulfur probably resulted from the reduction of sulfate to sulfite and then to free sulfur, or from the oxidation of hydrogen sulfide.

Strictly Autotrophic Sulfur Bacteria

The strictly autotrophic forms are found in the genus *Thiobacillus*. The cells are small, Gram-negative rods, nonmotile or motile. Energy is derived from the oxidation of incompletely oxidized sulfur compounds,

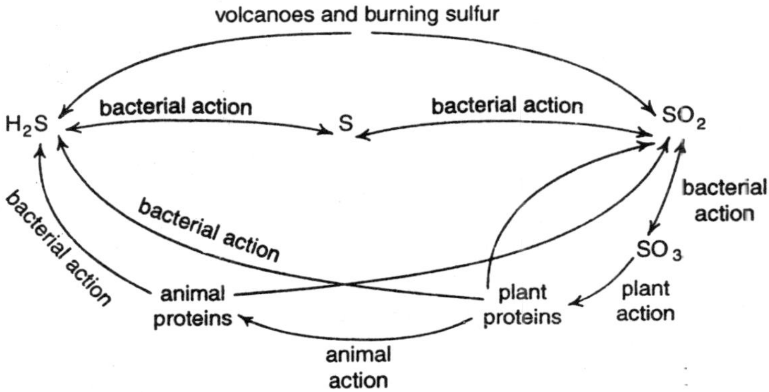

Fig. 5.5. The sulfur cycle.

principally from elemental sulfur and thiosulfate but in some cases also from sulfide, sulfite, and polythionates. The principal product of oxidation is sulfate, but sulfur is sometimes formed. The organisms grow under acid or alkaline conditions and derive carbon from carbon dioxide or from bicarbonate in solution. The strictly autotrophic species include *T. thioparus, T. neapolitanus, T. f errooxidans, T. concretivorus,* and *T. thiooxidans.* The first two can grow under alkaline conditions; the latter three, under strongly acid conditions. All species are aerobic, some being strictly so. Organisms may be cultivated in a mineral medium containing thiosulfate as the source of energy. In addition to thiosulfate, one species, *T. f errooxidans,* can oxidize ferrous iron compounds to ferric hydroxide for energy. The organisms become heavily encrusted with the reddishcolored compound.

Thiobacillus thiooxidans

This organism may be cultivated in an acid mineral medium containing free sulfur. Since it is a strict aerobe, growth is hastened if the medium is exposed in shallow layers. This species is of special interest in that it produces more acid from the oxidation of sulfur and continues to live in a more acid medium than any other living organism yet reported. The hydrogen-ion concentration of the medium increases to pH 0.6 and less. The cells actually grow in 5 to 7 per cent sulfuric acid.

$$2S + 3O_2 + 2H_2O \rightarrow 2H_2SO_4$$

Corrosion of Concrete and Sealing Mixtures

The thiobacilli have been shown to cause serious corrosion of concrete. Sufficient sulfuric acid is produced from the oxidation of sulfur or hydrogen sulfide to dissolve concrete. The species principally involved are *T. thioparus, T. concretivorus,* and *T. thiooxidans.*

Materials containing sulfur are used to seal the joints of cast-iron pipes. The molten mixture is poured into the joints, where it forms an impervious cement. Ordinarily this sealing material remains intact for years, but occasionally the sulfur cement breaks down under biological action. The principal organism producing this breakdown is *T. thiooxidans,* although other species are doubtless involved. The sulfur is first oxidized to sulfate, followed by the reduction of the sulfate to sulfide by sulfate-reducing bacteria.

Facultatively Autotrophic Sulfur Bacteria. A large number of sulfur organisms do not appear to be strict autotrophs. Because of their questionable nature, *they* are classified as facultatively autotrophic for convenience.

The organisms are commonly present in water containing dissolved hydrogen sulfide. The gas results from the decomposition of organic matter by saprophytic organisms. The bacteria may be present in such organic materials as decomposing seaweed, rock pools containing dead algae and other lower forms of plant life, stagnant woodland pools, and sewage. Sulfur bacteria have been found in sulfur hot springs and in sulfur mines. Some species are able to grow in water pipes and cause serious obstructions. Foul odors and tastes are produced after death and decomposition of the organisms.

The important facultative sulfur bacteria are found in the families Achromatiaceae, Beggiatoaceae, Chlorobacteriaceae, Thiobacteriaceae, and Thiorhodaceae.

Achromatiaceae

Large cells, spherical to ovoid or shortly cylindrical with hemispherical extremities. Division of cells by constriction in the middle. Do not possess photosynthetic pigments. In natural habitat, cells contain sulfur droplets and sometimes additional inclusions, such as spherules of calcium carbonate, Found in fresh water and marine environments.

Beggiatoaceae

Individual cells generally not visible without staining, occur in trichomes; within trichomes cells arranged in chains. Trichomes show a gliding motion when in contact with substrate; also show flexing movements. When grown in presence of hydrogen sulfide, trichomes contain sulfur globules. Do not possess chlorophyll and phycocyanin.

Found in both fresh water and marine environments containing hydrogen sulfide.

Chilorobacteriaceae

Green bacteria, usually small in size, occurring singly or in masses of various shapes and sizes, growing in environments containing high concentrations of hydrogen sulfide and exposed to light. As a rule, do not contain sulfur globules but frequently deposit elemental sulfur outside the cells. Contain chlorophylllike pigments. Capable of photosynthesis in presence of hydrogen sulfide; do not liberate oxygen.

Thiobacteriaceae

Coccoid, straight, or curved rods. Oxidize sulfur compounds and deposit free sulfur granules within or without the cells. Never filamentous. Colorless bacteria which are sometimes embedded in gelatinous pellicles or in gelatinous bladder-like colonies. Found where hydrogen sulfide occurs, or may oxidize free sulfur, thiosulfate, or related compounds.

Thiorhodaceae

Unicellular organisms, often developing as cell aggregates or families of variable size and shape. Single cells are spheres, ovoids, short rods, vibrios, spirals, long rods, or occasionally chains. Occur in nature in environments containing sulfides and require light for development. Produce a pigment system composed of green bacteriochlorophyll and yellow and red carotenoids. They appear as bluish violet, pale purple, brownish to deep red cell masses. Single cells usually appear to be unpigmented. Anaerobic or microaerophilic organisms with a photosynthetic metabolism in which carbon dioxide is reduced with the aid of special hydrogen donors without the liberation of molecular oxygen. Where these organisms are found in nature, hydrogen sulfide acts as a hydrogen donor, and sulfur accumulates as sulfur droplets in the cells. Probably all can utilize a number of organic substances in place of hydrogen sulfide as hydrogen donors for photosynthesis.

IRON BACTERIA

Iron is found in all living cells. In some species only minute amounts are present, but regardless of the quantitative aspects, the element is absolutely necessary for growth and well-being of all plant and animal cells.

Magnesium rather than iron is present in chlorophyll, the green coloring matter of plants. However, in the absence of iron, leaves do not become green. Also, if iron is withheld from a plant in which the chlorophyll is well developed, the color gradually fades to yellow.

Certain organisms found in water and soil are capable of taking up iron and accumulating it on the surfaces of their cells, where it is quickly oxidized to ferric hydroxide. The important iron bacteria are found in the orders Pseudomonadales and Chlamydobacteriales.

Strictly Autotrophic Iron Bacteria

Ferrobacillus f errooxidans is a strictly autotrophic iron bacillus. It utilizes CO_2 of the atmosphere as a source of carbon and derives its energy from the oxidation of ferrous iron to basic ferric sulfate or insoluble ferric hydroxide. The optimum reaction for growth is pH 3.5, and the optimum temperature is 20 to 25°C. The organism is indigenous to bituminous coal regions.

Nonautotrophic Iron Bacteria

A number of nonautotrophic species are found in iron-bearing waters. Such organisms are capable of utilizing both inorganic and organic matter.

Gallionella

Stalked bacteria, the long axis of the rod-shaped cells set at right angles to the axis of the stalks. Stalks are slender, twisted bands, dichotomously branched, composed of ferric hydroxide, completely dissolving in dilute hydrochloric acid. Multiplication by transverse binary fission. Young cells colorless, later becoming brown to rust red through the accumulation of iron. Twisted filaments easily identified since no other organism of a similar character has even been observed to take such a form. When the loops of the coils become encrusted with iron, the filament resembles a row of beads. Presence of a sheath not observed. Organisms attach themselves to pipes and cause extensive deposits of iron which interfere with the flow of water. For this reason, they are sometimes known as the water-pest bacteria. Organisms widely distributed in nature.

Sphaerotilus

Attached or free-floating, colorless trichomes showing false branching, though this may be rare in some species. Sheath shows a homogeneous structure. Sheath may become yellowish or brown with deposition of iron oxide. Deposition of iron dependent on environmental factors, not on physiological ability to store iron. Trichomes consist of rod-shaped or ellipsoidal cells surrounded by a firm sheath.

Multiplication occurs both by nonmotile conidia and by motile swarm cells, the latter with a subpolar tuft of flagella. Found in fresh water.

Leptothrix

Trichomes of cylindrical, colorless cells with a sheath at first thin and colorless, later thicker, ellow or brown, encrusted with iron or manganese oxide. Oxides may be dissolve by dilute acid, exposing the inner cells. If sheath contains manganese oxide, it does not dissolve completely in weak acids. Sheath shows an alveolar structure. Multiplication by cell division with individual cells occasionally slipping out of the sheath as reproductive cells. These are sometimes motile with a tuft of flagella. False branching may occur. Usually found in fresh water. Best known species, *L. ochracea,* is world-wide in distribution.

Toxothrix

Trichomes composed of cylindrical, colorless cells with a thin primary sheath; latter soon becomes impregnated with iron oxide. Trichomes lie loosely, longitudinally together, in slightly spirally twisted rolls. Continued repetition of this process leads to the development of a thick, secondary sheath from which parallel bundles may separate. False branching may occur. Sheaths do not completely dissolve in weak acids. Cells may slip out of sheath and become motile swarm spores. Found in fresh, iron-bearing waters.

Crenothrix

Trichomes attached to a firm substrate and swollen at the free end. Unbranched or show false branching. Sheaths surrounding trichomes are plainly visible, thin and colorless at the tip and encrusted with iron or manganese oxides at the base. Cells disk-shaped to cylindrical, dividing to produce spherical, nonmotile conidia of two types: micro- and macroconidia. Individual cells may slip out of sheath and form new trichomes. Found in stagnant and running waters which contain organic matter and iron salts. *C. polyspora* only species recognized.

Clonothrix

Attached trichomes showing false branching. Sheaths organic, encrusted with iron or manganese, broader at the base and tapering toward the tip. Cells colorless, cylindrical. Reproduction by spherical conidia formed in chains by transverse fission of cells; conidia formation acropetal, limited to short branches of the younger portions of the trichomes. Only species, *C. putealis,* widely distributed in rivers and streams with gravelly, manganese-bearing bottoms; also found in water works. and pipe lines, where it may cause technical difficulties. Ferruginous waters usually show the presence of a yellowish- or reddish-colored slime on the stream bottom. The color is due to the ferentiation deposition of iron in the outer sheaths of the filaments. The accumulation of iron and its oxidation to ferric hydroxide result in the formation of a hard and inelastic membrane, which eventually leads to the death of the organisms. Old filaments show a higher iron content than do young filaments. In some cases, young cells are completely lacking in a deposition of iron in their sheaths. The iron hydroxide may be removed by the application of dilute hydrochloric acid, after which the outer membrane becomes visible.

6

GROWTH OF BACTERIA

In this chapter we shall be concerned with how bacteria are grown, methods by which we can measure their growth, and how a single species of bacteria can be isolated from a mixture of organisms. Various methods that can be used for the physical or chemical sterilization of a given area or object are also discussed. All organisms, whether procaryotic or eucaryotic, require food to live and grow. How an organism assimilates its fold is called its nutrition, and the specific cellular requirements are its nutrients.

The cell also requires a source of energy for the synthesis of cellular constituents and for other life processes, such as motility and active transport. The major metabolic pathways by which bacteria acquire and convert their nutrients into energy and the metabolic products required for growth are discussed.

GROWTH OF BACTERIAL POPULATIONS

Unlike most higher organisms, bacteria are unicellular, which means tat during normal growth the increasing in the population is simply related to the growth of any cell in that population. A bacterial cell in a satisfactory growth medium will grow larger by means of a balanced increase in all of its constituents. After it has doubled in size, cell division can occur, resulting in two identical cells, each of which is capable of growing and dividing at the same rate as the original cell. This means that the number of cells in the population increases in a exponential or logarithmic mannér, that is, each doubling of the population occurs in the same time period, but the number of cells that participate in each division doubles for each succeeding generation.

Normal Bacterial Growth Curve

By its very nature, bacterial reproduction by binary fission results in doubling of the number of viable bacterial cells. Therefore, during active bacterial growth, the size of the microbial population is continuously doubling. Once cell division begins it proceeds exponentially as long as growth conditions permit, with one cell dividing to form two, each of these cells dividing so that four cells form, and so forth in a geometric progression. The time required to achieve a doubling of the population size, known as the generation time or doubling time, is the unit of measure of microbial growth rate. The generation time for bacteria can be expressed mathematically as

$$g = t/3.3\ (\log B_t - \log B_0)$$

where g is the generation time, log B_t is the logarithm to the base 10 of the number of bacteria at time t, log B_0 is the logarithm to the base 10 of the number of bacteria at the starting time, and t is the time period of growth. By determining cell numbers during the period of active cell division, the generation time can be estimated. A

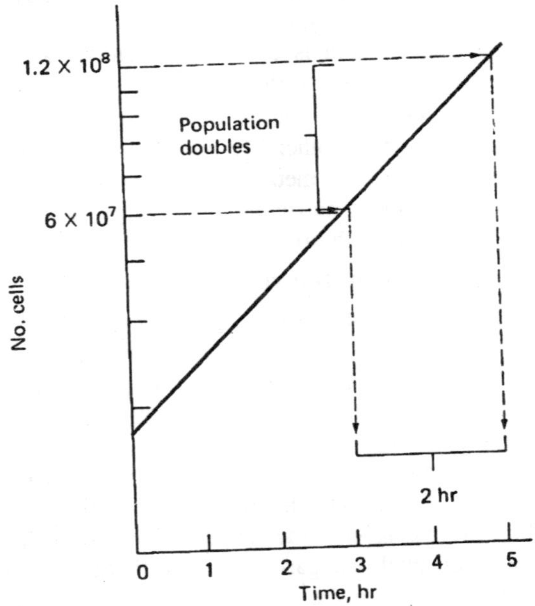

Doubling time = 2 hr

Fig. 6.1. The estimation of the generation time of a bacterium based on observed celi numbers during a period of active division.

bacterium such as *E. coli* can have a generation time as short as 20 minutes under optimal conditions. Considering a bacterium with a 20-minute generation time, one cell would multiply to 1000 cells in 3.3 hours and to 1,000,000 cells in 6.6 hours.

Table 6.1. Growth rates for some representative bacteria under optimal conditions

Organism	*Temperature (°C)*	*Generation Time (min)*
Bacillus stearothermophilus	60	11
Escherichia coli	37	20
Bacillus subtilis	37	27
Bacillus mycoides	37	28
Staphyolococcus aureus	37	28
Streptococcus lactis	37	30
Pseudomonas putida	30	45
Lactobacillus acidophilus	37	75
Vibrio marinus	15	80
Mycobacterium tuberculosis	37	360
Rhizobium japonicum	25	400
Nostoc japonicum	25	570
Anabaena cylindrica	25	840
Treponema pallidum	37	1980

When a bacterium is inoculated into a new culture medium, it exhibits a characteristic growth curve. The normal growth curve of bacteria has four phases, the lag phase, the log or exponential growth

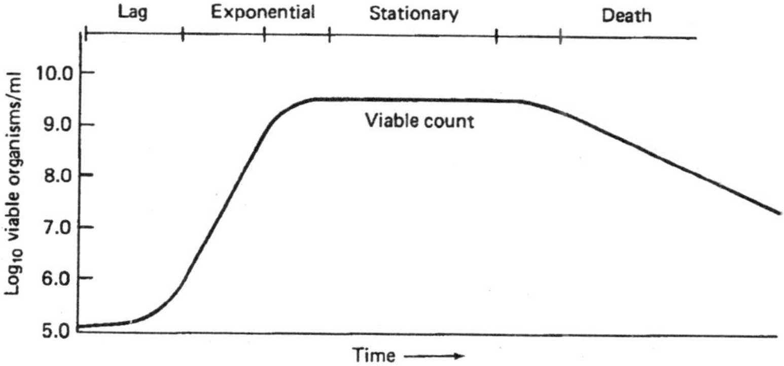

Fig. 6.2. The normal bacterial growth curve has four stages: lag, exponential, stationary and death.

phase, the stationary phase, and the death phase. During the lag phase there is no increase in cell numbers. Rather, during this phase the bacteria are preparing for reproduction, synthesizing DNA and various enzymes needed for cell division. During the log phase of growth, so-named because the logarithm of the bacterial biomass increases linearly with time, bacterial reproduction occurs at a maximal rate for the specific set of growth conditions. It is during this period that the generation time of the bacterium is determined.

If a bacterial culture in the exponential growth phase is inoculated into an identical fresh medium, the lag phase is bypassed, and exponential growth continues. This occurs because bacteria are already actively carrying out the metabolism necessary for continued growth. If, however, the chemical composition of the new medium differs significantly from the original growth medium, the bacteria do go through a lag phase before entering the logarithmic growth phase when they synthesize the enzymes needed for growth in the new medium.

However, if the bacterium is not transferred to a new medium and no fresh nutrients are added, the stationary growth phase eventually is reached, and there is no further net increase in bacterial cell numbers. During the stationary phase the growth rate is exactly equal to the death rate. A bacterial population may reach stationary growth when a required nutrient is exhausted, when inhibitory end products accumulate, or when physical conditions are restricting. The duration of he stationary phase varies; some bacteria exhibit a very long stationary phase.

Eventually the number of viable bacterial cells begins to decline, signaling the onset of the death phase. The kinetics of bacterial death follows the same exponential kinetics as the logarithmic growth phase. This is true because the death phase really represents the result of the inability of the bacteria to carry out further reproduction. The rate of the death phase need not, however, be equal to the rate of growth during the exponential phase. The rate of death is proportional to the number of the survivors and can be expressed as where d is the death rate, t is the time period, log B_0 is the logarithm of the number of bacteria at the starting time, and log B_t is the logarithm of the survivors at time t.

Batch and Continuous Growth

The normal bacterial growth curve is characteristic of bacteria in batch culture, that is, under conditions when a fresh medium is simply inoculated with a bacterium. A flask containing a liquid nutrient medium inoculated with a bacterium such as *E. coli* is an example of a batch culture. In batch culture growth nutrients are expended and metabolic products accumulate in the closed environment. The batch culture models situations such as occur when a canned food product is contaminated with a bacterium.

Bacteria may also be grown in continuous culture in a chemostat, where nutrients are supplied and end products continuously removed so that the exponential growth phase is maintained. Because end products do not accumulate and nutrients are not completely expended, the bacteria never reach stationary phase. The chemostat is a device in which a liquid medium is continuously fed into the bacterial culture. The liquid medium contains some nutrient in growth limiting concentration, and the concentration of the limiting nutrient in the growth medium determines the rate of bacterial growth. Even though bacteria are continuously reproducing, a number of bacterial cells are continuously being washed out and removed from the culture vessel.

Bacterial Growth on Solid Media

The development of bacterial colonies on solid growth media follows the basic normal growth curve. However, the dividing cells do not disperse; hence the population is densely packed. Under these conditions nutrients rapidly become limiting at the center of the colony, and microorganisms in this area rapidly reach stationary phase. At the periphery of the colony, cells can continue to grow exponentially even while those at the center of the colony are in the death phase. Bacterial colonies generally do not extend indefinitely across the surface of the media but have a well-defined edge. Therefore, individual well-isolated colonies develop from the growth of individual bacterial cells. The fact that the bacteria have reproduced asexually by binary fission means that, barring mutation, at the bacteria in the colony should be genetically identical; that is, each colony contains a clone of cells derived from a single parental cell.

Cytological Approach

Cells that are growing are destined to divide. Cell division takes place at different intervals depending on the species and environmental conditions. Within a short period, often as small as 20 minutes, a bacterium can create a complete duplicate of itself, which is in turn capable of duplicating. This process requires the synthesis of the millions of parts of the parent cell and their organization into the various homo-and heteropolymers which are the functional units of the cell.

In order to understand bacterial growth it is necessary to have some knowledge of the temporal sequence of events during a single cell cycle and an understanding of how these biochemical events are interrelated. Bacteria do not exhibit the characteristic cell-division cycle observed in eucaryotic organisms. Whereas in eucaryotic organisms DNA synthesis is confined to a single phase of the cell cycle, the S phase, in bacteria that are growing exponentially DNA synthesis occurs throughout virtually the entire division cycle. Also, in bacteria the duplication sequences do not necessarily follow one after the other, but instead they overlap, the amount overlapping depending on the growth medium. Bacterial duplication begins with the initiation of DNA replication, followed by the complete duplication and separation of the DNA replicas, and formation of a septum near the equatorial plane so that two cells, each one containing half the DNA content of the mother, are formed.

Techniques that achieve synchrony in a dividing population in a metabolically undisturbed state have proved invaluable in studies of the bacterial division cycle. Autoradiography and marker-frequency analysis have been used for the study of particular aspects of the division cycles in asynchronous cultures; morphologic details related to the division process have been provided by examination of serial ultrathin sections with the electron microscope.

Bacterial Cell Division

Genetic information necessary for duplication of the bacterial cell is contained within the DNA-containing structures, the bacterial chromosomes. The number of chromosomes in a cell; and the extent of their replication are determined by the physiologic state and age of the cell. Chromosomes of cells in very slowly growing cultures appear to have but a single replication point. Observations of gaps in DNA synthesis in these cells indicate that chromosome replication occupies only part of the cycle. Cells in rapidly growing cultures, on the other hand, appear to contain multiple replication points. Because replication occurs simultaneously at several points along these chromosomes, the entire chromosome can be replicated in a fraction of the time that would be required in there were but a single replication fork, and genetic material can be duplicated at rates that otherwise could not be attained.

Initiation of Chromosome Replication

During division in order to insure the segregation of equal portions of the genetic material among the progeny, chromosome replication and cell division are coordinated in such a manner that the frequency of initiation of replication determines the rate of cell division. As presently conceptualized, a duplication sequence of bacteria begins with the initiation of chromosome synthesis at a unique site, the replicator site of the chromosome. Initiation of chromosome replication requires proteins synthesis and depends on the achievement of a critical cellular mass/DNA ratio. Chromosome replication begins when a fixed amount of protein initiator substances have accumulated regardless of the position of other replication points on the chromosome. Thus, the rate of cell division in a given culture medium is determined by the time required for accumulation of the initiator in that medium. Initiation is not influenced by the presence, absence, position, or rate of movement of replication points on the chromosome.

Following initiation the replication point progresses sequentially along the chromosome, and initiator proteins continue to accumulate. After a time, depending upon the nutritional conditions, another threshold quantity of initiator will have accumulated. This results in the initiation of new rounds of replication, and the movement of new replication points along the chromosome. In cells growing under conditions in which the time for formation of initiator is longer that the time required for a replication point to traverse the chromosome, a new round the replication is not initiated until after the previous replication point has reached the chromosome terminus. In this case, there is a period devoid of DNA synthesis between rounds of chromosome replication. Conversely, in cells growing in media, in which the interinitiation time is shorter than that required for a round of replication, new rounds are initiated before the previous replication point has reached the terminus. In this situation, cells contain chromosomes with three or more replication points, as have been observed in rapidly growing cells. Helmstetter's model of DNA replication is based on the constancy at all growth rates of two periods: (1) the interval between the end of a round of DNA replication and cell division and (2) the time required for a replication point to traverse the genome. The first of these periods could be considered to be analogous to the G_2 period in eucaryotic cells; the second period, analogous to the S period, is the time required for the synthesis of DNA.

Chromosome Duplication

The replication of bacterial DNA begins by the gradual separation of the two strands at the replicator site to form a Y-shaped locus, the replication fork, which represents a growing point. The replicating enzyme system attaches to the DNA at this site and moves along the entire chromosome replicating the parent DNA as the double helix slowly rotates and unwinds. After a replication point reaches the terminus of a chromosome, the two newly replicated sister chromosomes separate form each other. Although the process has been referred to as "nuclear division," it is not strictly analogous to nuclear division in eucaryotic organisms since the chromosomes separate gradually throughout the entire period of replication.

Role of Cell Membrane in Chromosome Separation

Electron microscopy has failed to show any bacterial structure equivalent to the mitotic apparatus of higher cells. It is evident, however, that there must be some mechanism which ensures equitable separation of chromosomes after each replication. The replicon method proposed by Jacob and his associates provides an explanation of how this might be achieved. The model suggests that the cellular DNA is attached to the bacterial membrane where replication occurs by passage of the DNA through an enzyme complex in the membrane. Separation and equipartition of the DNA copies results from the synthesis of new cell membrane in an annular region between attachment points.

This model is supported by both biochemical and morphologic criteria. Biochemical analyses have demonstrated that in gently lysed preparations of bacteria the actively replicating portion of the DNA is associated with a membrane fraction. Electron microscopic studies have provided a morphologic basis and have shown that in gram-positive organism extensions of the cytoplasmic membrane, the mesosomes, are in contact with the bacterial chromosome. Serial sections through dividing cells have indicated that in the course of replication the mesosomes with thin attached chromosomes split in two and move apart, each carrying with it one of the two sister chromosomes. Evidence that the separation of the mesosomes is achieved by the synthesis of new membrane in the region between them was obtained by following the distribution of a specific marker, potassium tellurite, in the membrane during growth. Potassium tellurite is deposited selectively I the reduced form at the oxidation-reduction sites of the membrane as crystals that are visible in the electron microscope. The observed distribution of crystals supports the hypothesis that new membrane is generated in the region of the point of attachment of the chromosome to the membrane.

In gram-negative bacteria where mesosomal formation is minimal, the chromosome is either attached to the membrane without the mediation of mesosomes, or it is attached to their base. There appears to be no unique membrane attachment site on the *E. coli* genome; either the genome is attached to the membrane at random, or to a number of fixed or changing attachment points.

Terminal Steps in Cell Division

Following the separation of the two daughter chromosomes the cell initiates synthesis of a transverse septum. The temporal relationship between replication of the chromosome and synthesis of the transverse septum is closely regulated. The completion of a round of replication together with a period of protein synthesis which normally occurs concurrent with chromosome replication are both required for cell division. Cell division occurs at a fixed time after the termination of a round of DNA replication. This time period represents the time required for the expression of the many special reactions involved in septum formation, membrane growth, and wall synthesis.

The importance of the coupling between DNA replication and cell division has been emphasized by the finding of a number of mutants in which the regulation of cell division is disrupted. Many of these are conditional lethal mutants that grow normally at 30°C but are unable to form colonies at 41°C. In some of these mutants specific differences in membrane proteins have been observed, some of which correlate well with an interference of septum formation and inhibition of cell division.

Septum formation

Compartmentalization of physiologic division of the cell occurs prior to the actual physical separation of the daughter cells. This is the time at which the cytoplasmic events at one end of the cell become independent of cytoplasmic events occurring at the opposite end. The end of a round of replication occurs well in advance of this compartmentalization.

This stage of growth is characterized by the development of a weak septum followed by the formation of a strong cross-wall. As emphasized previously, the position and timing of septum formation is crucial if each daughter cell is to receive a full compliment of the parental material. The process of septum formation is poorly understood and little is known of the physiologic events which periodically initiate it. Although it is generally agreed that the completion of chromosome replication is necessary for division, other factors such as the presence of polyamines and heat-labile proteins also play an important role.

Cell-wall growth

During balanced growth a greater surface area is normally required to cover a greater volume. In several gram-positive species, cell-wall growth appears to be limited to discrete areas at the equator. These areas have been located in a number of gram-positive organisms by staining the cell surface with fluorescent antibodies and observing the location of the label during subsequent growth. In gram-negative organisms, however, although zonal growth of the murein component also occurs, the lipopolysaccharide of the cell envelope is probably added by diffuse intercalation during cell elongation.

The nascent cross-wall of the growing cell originates within the area enclosed by a mesosome. Subsequent inward growth of the wall is preceded by a corresponding inward proliferation of the mesosome which is formed by a concentric infolding of the plasma membrane. As the mesosome, followed by the growing transverse septum, moves toward the center of the cell, the membranes at its base continue to form around the newly synthesized cell wall.

A notch appears in the external wall directly above the developing cross wall, accompanied by the formation of two new raised bands adjacent to the notch. These bands appear to mark the boundary between new and old wall. Both peripheral wall elongation and centripetal cross-wall extension apparently result from biosynthetic activity in the vicinity of the leading edge of the cross wall. An interplay of murein hydrolyzing and murein-synthesizing enzymes has been invoked to explain the sequence of interrelated events associated with the enlargement of the cell wall. A closed molecule like the structural component of the cell wall, the murein sacculus, can only be modified when covalent bonds are split. The hydrolyases would provide both space and acceptor sites for incoming subunits which in a second step would be condensed into the preexistent murein. The sites at which murein hydrolases are active correspond to the zone where murein is synthesized.

Separation of bacteria

Separation of the two daughter cells takes place some time after the completion of the transverse septum. The separation begins with the appearance of a constriction between the subsequent daughter cells and a thickening of the cell wall in the constriction region. Further thickening of the cell wall accompanies constriction between the cells until the pole of the cell assumes its characteristic rounded shape. The normal thickness of the cell wall, however, is achieved only after constriction is complete and the two daughter cells have completely separated. Separation of the daughter cells is mediated by murein hydrolases which are localized at the site of septum formation, and which are latent in nondividing cells.

Asymmetrical Cell Division

In addition to the above method of cell division which insures that each daughter cell receives a nucleus following cell division, an unusual form of division has been observed in several mutant strains. In a mutant of *E. coli*, small anucleate cell-forms are produced during exponential growth. These small forms are called minicells and are formed near one pole of the parent cell. Since they lack DNA, they do not divide further and are unable to manufacture any polymer whose synthesis is DNA dependent.

A series of temperature-sensitive mutants of *E. coli K*12 has also been isolated which grow normally at 30°C but cease DNA synthesis at 40°C after a small portion of the chromosome has been replicated. Most of these mutants are unable to form a septum so that the cells continue to elongate and to produce long filamentous cells without septa. However, a small number of these thermosensitive mutants can still form septa and undergo cell division. The septa in these mutants are not formed near the equatorial plane as in the wild-type parent but to one side of the DNA, so that of the two daughter cells formed, one contains all of the DNA of the mother bacterium and the other none. This type of anucleate cell, however, differs from the minicell in that it is about the same size as a normal wild-type cell.

Measurement of Growth

Isolation of pure cultures

The first step in the measurement of the growth properties of a bacterium is to ensure that one is dealing with a pure culture. This is of special concern in studies of bacteria from clinical samples that contain large numbers of different cells, each of which can respond differently to changes in the growth medium as well as influence the growth of other bacteria in the population. Pure cultures are usually isolated by spreading out a sample of cells on the surface of a suitable medium. A variety of growth media has been developed for the isolation or identification of specific types of bacteria. Some media are selective, meaning that they contain ingredients that inhibit the growth of certain bacteria. An example of this is the inclusion of bile salts and basic dyes in a medium used for the isolation of enteric bacteria; these reagents are able to penetrate the gram-positive cell wall and thus prevent the growth of most gram-positive bacteria. Other media re termed differential, meaning that they permit the differentiation of organisms on the basis of some biochemical property. For example, media used for the isolation of enteric bacteria incorporate lactose and an acid—base indicator into a peptone containing medium. As a result, those organisms that are able to utilize lectose and produce acid from it can be readily distinguished from other enteric bacteria, which grow on the peptone present in the medium and do not produce acid from these peptides.

Almost all media used for the purification of bacteria are solidified by the incorporation of agar, a complex polysaccharide isolated from seaweed. It has the desirable properties of melting at about 95°C, but forming a solid gel at about 45°C. Thus, it can be kept molten at temperatures above 50°C and, if desired, mixed with other heat-labile ingredients prior to being poured into suitable vessels and allowed to harden. Fortunately, few bacteria degrade agar.

Bacteria will grow when placed on an agar surface of a suitable medium either by pouring or by spreading with a wire loop. Wherever a single cell has been deposited, a visible colony will arise containing 10^6 to 10^9 bacteria, all of which are the direct progeny of the original cell at that site. An isolated colony thereby provides the pure culture needed for subsequent studies of its growth and drug-resistance properties. Good practice suggests that such a single colony be restreaked on a similar medium to ensure the absence of any contaminating bacteria.

Determination of bacterial mass

To measure bacterial growth and the effect of antibiotics on a particular bacterial strain, it is necessary to determine both the mass of the cells and the number of living cells in the sample. No single method allows both measurements. Cell mass can be determined from the dry weight, although this requires large samples and considerable time. Somewhat more rapid methods involve measurement of cell volume or the total amount of cell nitrogen of protein, using calorimetric assays. By far, however, the simplest and most frequently employed technique is to measure the optical density or turbidity of a liquid culture. This assay, which can be readily automated, is based on the fact that light scattering by particles is related to the number, size, and mass of the particles. The determination of certain metabolic activities, such as dye reduction by the electron transport components or the release of radioactive CO_2 from a labelled substrate, are also being explored, since such procedures are far more sensitive than those based on turbidity.

Determination of cell number

The total number of cells, both living and dead, can be determined microscopically by using a special glass slide, called a Petroff-Hauser Counter, which has a graduated chamber of known volume. For those who require a frequent determination of cell number, an electronic particle counter, known as a Coulter counter, provides both the number of particles and their size distributions.

The number of living cells in a given sample can be determined by plating a dilution of the original culture onto the surface of an agar medium that has solidified in a Petri dish. The diluted sample may also be mixed with the melted agar medium before pouring it into a Petri dish to harden. The number of colonies that arise after 1 or 2 days' incubation are counted and multiplied by the dilution factor to yield the number of viable cells per milliliter in the original sample. Figure illustrates such a dilution scheme in which the sample was diluted by a factor of 10^{-6}, and 0.1 ml was plated onto the surface of the plate. In this case, if 90 colonies were found on the plate after incubation, one could calculate that there were 90×10^{6} cells per 0.1 ml, or 90×10^{7} viable cells per ml of the original sample. Clinical samples expected to have few cells, such as urine or clear water, can be concentrated by centrifugation or filtration rather than diluted prior to plating.

Phases of Growth

Under favourable conditions, most bacteria are able to reproduce rapidly. The generation tine is the time required for the population to double. Examples near both ends of the spectrum would include *Clostridium perfringems*, which has an optimal generation time as short as 8 to 10 minutes, and *Mycobacterium tuberculosis*, in which it is about 15 to 20 hours. As one might expect, the generation time varies according to the growth medium and the species. Because most organisms have generation times in the range of 30 to 120 minutes, bacterial populations can quickly become quite large. For example, one *Escherichia coli* cell with a 30-minute doubling time will grow overnight into more than one billion (10^{9}) cells. Assuming division at the same rate, it would require only an additional 37 hours for the progeny of that single cell to cover the United States to a depth of 1 meter. These simple examples stress the rapidity of bacterial growth and the necessity for speed in determining the number of cells in a sample. When cells are inoculated into a growth medium, the increase in their number and mass can be plotted on a logarithmic scale as a function of time. Such a growth curve reveals four major phases.

Lag phase

When bacteria are inoculated into fresh medium, reproduction does not usually begin immediately. During this lag phase, the cell mass and size begin to increase long before cell numbers increase. It is during this phase that the synthesis of enzymes and other macromolecules needed for growth in the new medium occurs. The length of the lag phase depends on the kind of bacteria, the age and size of the inoculum, the nature of the medium from which they were taken, and the nutrients present in the new medium.

Log or exponential growth phase

During the exponential growth phase, cell mass and numbers increase in a logarithmic manner with a constant generation time. This is a period of balanced growth during which all cellular constituents are increasing in constant proportion. The rate of cell division, is of course, dependent on the type of organism, the nature of the medium, the temperature, and, for aerobic organisms, the rate of aeration.

The generation time can be determined during the log phase, because during this phase, both cell number and mass increase in a coordinate manner. Thus, measurements of either parameter can be used for this determination, and since each generation results in a doubling of the cell population, the following equation applies:

$$B_t = B_0 \times 2^n$$

where B_t and B_0 are the number of bacteria or their mass at time t and at the initial time 0, respectively, and n is the number of generations that have occurred between time 0 and time t. Taking the logarithm (to the base 10) of both sides of the equation yields.

$$\text{Log } B_t = \log B_0 + n \log 2$$

Since $n = t/g$, where t is the time and g is the generation time in minutes per doubling, the equation above can be rearranged to yield

$$\text{Log } B_t - \log B_0 = \log 2 \times t/g$$

This equation can be rearrange to solve for any unknown when one is given three of the four parameters (log 2=0.301). A frequently employed function is called the "specific growth rate," which is given the symbol, μ, and is in units of doublings per hour.

$$\mu = \ln 2/g = (\text{In } B_t - \ln B_0)/t$$

Stationary phase

Logarithmic growth eventually slows because of accumulation of waste products, exhaustion of nutrients, change in pH, or a decrease in oxygen tension. The population then enters the stationary phase in which the number of viable cells remains approximately constant. Usually, there is a division of some cells and the death of others. It is worth noting that entry into the stationary phase or even starvation for a required nutrient does not result in the killing of most bacteria.

Death phase

Eventually the rate of death exceeds the rate of reproduction, and the number of viable cells declines. The length of tie before all cells have died differs markedly for various organisms. Some species have a very short death phase, whereas for others it may take weeks before all of the cells in a culture have died. During this phase, cell often assume unusual shapes, making it difficult to recognize the bacteria in old cultures.

DIFFERENTIATION IN BACTERIAL CELLS

Sporulation and Germination

One of the most unusual properties of certain bacteria is the ability to form endospores. At some point in the vegetative cell cycle of spore-forming organisms growth is arrested and the cell undergoes progressive physiologic and morphologic changes which result in the formation of an endospore. A spore is a dormant structure capable of surviving for prolonged periods, but endowed with the capacity to reestablish the vegetative stage of growth under appropriate environmental conditions. The processes involved in sporulation as well as the breaking of the spore's dormancy and subsequent emergence of a vegetative cell, represent primitive examples of unicellular differentiation. Studies of these phenomena provide model systems for understanding the mechanisms by which sequential morphologic and biochemical events are ordered and controlled in a differentiating system.

Properties of Endospores

Endospores are formed by organisms of the genera *Bacillus*, *Clostridium*, and *Sporosarcina* during the stationary phase of growth after certain essential nutrients in the medium have been exhausted. A single spore is produced within a vegetative cell and differs from it in its morphology and composition, increased resistance to adverse environments, and absence of metabolic activity, Although the thermoresistance of spores has been the major concern of medical microbiologists because of the problems associated with the control of infection caused by spore-producing organisms, spores also have an increased resistance to the lethal effects of desiccation, freezing, radiation, and deleterious chemicals. The primary selective value of the spore, however, lies in its longevity in the soil coupled with its ability to germinate under the proper environmental conditions.

Chemical basis of spore resistance

Resistance to various chemical and physical agents appears at different stages during sporulation, concomitant with changes in the physicochemical composition of the sporulating cell. The resistance to radiation, drying, and chemical agents appears after the cell becomes refractile and depends at least in part upon the properties of the cystine-rich spore coat proteins. Thermal resistance is probably due to the very low content of water, which renders the proteins and nucleic acids more resistant to denaturation. This dehydration occurs late in sporulation at the time of cortex formation and at the time that the spore first appears as a refractile object. The massive synthesis of dipicolinic acid during this stage may be responsible for the dehydration. Dipicolinic acid, a chelating agent, is a spore-specific component present in high concentrations in all bacterial endospores. It is present in the cortex of the mature spore as the calcium salt and accounts for 5 to 15 percent of its dry weight. Both calcium and dipicolinic acid are required for heat resistance.

Metabolic Differences Between Vegetative Cells and Spores

As a bacterial cell passes from the vegetative state to the sporulating state, it undergoes dramatic changes in its morphology and physiology. During sporulation certain compounds such as poly-b-hydroxybutyrate accumulate and are later utilized. Extensive turnover of macromolecules occurs, and some enzyme levels change drastically. New and characteristic spore structures are synthesized, while previously existing structures and degraded.

The pools of small molecules found in spores are distinctly different from those of vegetative cells. There is an accumulation of dipicolinic acid, divalent cations, and high levels of L-glutamic acid. The predominant component of the drastically reduced acid-soluble phosphate pool is 3-phosphoglyceric acid, which constitutes as much as 75 percent of the acid-soluble phosphate in the spore, but no more than 5 percent of the acid-soluble phosphate of the vegetative cell. Spores contain very low levels of ATP, whereas ATP, is a major component of vegetative cells.

During sporulation many differences occur both qualitatively and quantitatively in the pattern of enzyme activities. Some of these differences are associated with mechanisms related to spore formation; some are components of the spore itself. The heat-stable catalase of spores is immunologically distinct from the vegetative cell enzyme; glucose dehydrogenase, a ribonuclease, and spore lytic enzymes are present only in the spore and absent form the vegetative cells. These enzymes as well as the spore coat proteins represent spore-specific gene products.

Initiation of Sporulation

Sporulation is a response to nutritional deprivation. Although limitation of any one of a variety of nutrients can initiate sporulation, the most pronounced effect is exerted by the carbon and nitrogen sources available. The regulation of spore formation in the bacterial cell is negative; the cell makes an inhibitor or repressor from some ingredient in the medium, which prevents the initiation of sporulation. When this ingredient is exhausted, inhibition is released, and the cells sporulate. A metabolizable supply of both carbon and nitrogen are required to inhibit sporulation. If there is a deficiency in the supply of either of these, inhibition will be relieved and sporulation initiated.

Biochemistry of Sporulation

At some period in the development of a cell, the metabolism is irreversibly channeled in the direction of sporulation. There is no single point of commitment for the sporulation process as a whole, but a separate point of commitment for each new specific macromolecule. An examination of thin sections of sporulating cells in the electron microscope has shown that an ordered series of cytologic and structural changes accompanies the physiologic changes associated with sporulation. The morphologic changes observed in *B. subtilis* have been presented as a time scale divided into seven sequently appearing stages. The process is times from the end of exponential growth at t_0 and at hourly periods after this.

Stage 0. This represents the state of the cell at the end of exponential growth at a time when each vegetative cell contains two chromosomes.

Stage I. This stage is characterized by the formation from the more disperse nuclear bodies of a broad axial chromatin filament, which occupies the center of the cell. During this stage the most characteristic events are the production and excretion into the medium of antibiotics and various exoenzymes. The importance of antibiotic and protease production in the formation of spores is emphasized by the finding that mutants unable to produce these substances are almost invariably also asporogenous. Protease's probably play a role in thc intracellular turnover of protein by the correlation between sporulation and antibiotic production is more difficult to explain. The importance of antibiotic production per se to the sporulation process may result from the fact that it is merely one of an ordered sequence of biochemical events, some essential and some not to the sporulation process.

Stage II. This stage is characterized by the formation of a forespore septum near one pole of the cell. Septum formation follows the separation of the two discrete chromatin bodies in the axial filament and movement of one of them to the end of the cell, resulting in the segregation of the nuclear material into two compartments, the mother cell and the forespore units. The forespore septum is formed *de novo* and does not result from the rearrangement of pre-existing membrance. During this period there is a marked increase of alanine dehydrogenase, a component important in germination.

Stage III. During this stage the spore protoplast in formed as a consequences of the unidirectional growth of the cytoplasmic membrane of the sporangium. Among the enzymes whose activity is markedly increased during this stage are glucose dehydrogenase and aconitase. Aconitase is an example of an enzyme appearing during the early stages of sporogenesis which is, like sporulation itself, subject to catabolite repression. With the exhaustion of glucose and glutamate at the end of vegetative growth repression is relieved, aconitase is formed, and functional tricarboxylic-acid glyoxylic-acid cycles produced. Heat-resistant catalase and alkaline phosphatase also appear during Stage III.

Stage IV. Cortex formation occurs during this stage at a time when the spores appear for the first time as refractile objects. Metabolic activity of the forespore at the end of State IV is considerably less than that of a vegetative cell, and beyond this stage the forespore probably contributes little to its own development. The activity of both ribosidase and adenosine deaminase increases during Stage IV and the synthesis of dipicolinic acid begins.

Stage V. This is the period of coat formation. Deposition of spore coat protein and the incorporation of cystine lead to a completed coat structure.

Stage VI. Spore maturation occurs during this stage resulting in the fine structure of the cytoplasm of the spore protoplast which makes it more homogeneous and electrondense. There is continued synthesis of or modification of cortical peptidoglycan, and uptake of dipicolinic acid and calcium. During Stage VI alanine racemase, an enzyme important in spore germination, is formed, and a number of ill-defined changes occur which make the spores resistant to organic solvents and to heat.

Stage VII. The mature endospore is liberated from the mother cell. This is initiated by a lytic enzyme synthesized or activated subsequent to maturation or the endospore.

Protein Synthesis during Sporulation

With the appearance of the forespore the cell becomes a compartmentalized structure. In the compartment developing into the forespore, DNA replication is inhibited and the substances synthesized are spore-specific. In the second compartment, the sporangium, synthesis of vegetative DNA continues and products of the vegetative cell continue to be produced. Thus, in the sporulating cell both the vegetative and spore genomes are transcribed. The mRNA contains some of the same mRNA molecules as depressed cells transferred from a rich medium to a poor one, but in addition molecules specific to sporulation. There is no evidence that the mRNA produced late in sporulation is stable and retained in the spore for protein synthesis during outgrowth.

Role of RNA polymerase

The changes that occur during sporulation result from the turn-off of certain vegetative genes and the expression of new classes of genes. Ribosomal RNA genes are not transcribed in sporulating cells. The turn-off, however, of ribosomal RNA genes during sporulation is brought about by a mechanism different from that responsible for stringent control of ribosomal RNA synthesis. The most attractive model which has been proposed suggests that early in passing from the vegetative to the sporulating state, the template specificity of the RNA polymerase is altered. The vegetative σ factor does not function in sporulating cells while the core enzyme, the machinery for polymerizing RNA, is conserved. The reason for failure of the s factor to function may possibly be explained by differences observed between the β subunits in vegetative and sporulation RNA polymerase cores. The two β polypeptides of vegetative RNA polymerase have a molecular weight of 155,000 daltons. In sporulation RNA polymerase, one of these β subunits has a molecular weight of only 110,000 daltons. Although this modification of vegetative RNA polymerase is critical to sporulation and may be the event responsible for the turn-off of some vegetative genes during sporulation, a more precise explanation is still needed for the mechanism by which the transcription of new genes is directed. The spore genome is present in vegetative cells but is repressed under conditions of rapid growth. At the end of exponential growth the level of intracellular catabolites is altered, "and the gene governing the synthesis of a "sporulation initiation factor" is derepressed. The nature of this "sporulation initiation factor," however, remains to be clarified.

Germination and Outgrowth

Simultaneous structural and physiologic changes are also manifested during the transformation of a dormant spore into a vegetative cell. The process of spore germination consists of three sequential phase : (1) an activation state which conditions the spore to germinate in a suitable environment; (2) a germination stage, during which the characteristic properties of the dormant spore are lost; and (3) an outgrowth stage during which the spore is converted into a new vegetative cell.

Activation is a reversible process which conditions the spore for germination. Spores do not germinate or germinate very slowly unless activated either by heat or various chemical treatments. Activation probably involves a reversible denaturation of a specific macromolecule. Germination is an irreversible process triggered by the exposure of activated spores to specific stimulants such as amino acids, nucleosides, and glucose. This is the stage during which the dormant stage is ended. During the early stages of germination there is a loss of refractility, swelling of the cortex, and appearance of fine nuclear fibrils. Accompanying these changes is a loss of resistance to deleterious physical and chemical agents, increase in the sulphydryl level of the spore, a release of spore components, and an increase in metabolic activity. The germination of spores is not inhibited by antibiotics which perturb protein and nucleic acid synthesis, indicating that the enzyme responsible for germination are already present in the spore.

During outgrowth there is *de novo* synthesis of proteins and structural components which are characteristic of vegetative cells. During this stage the spore core membrane develops into the cell wall of the vegetative cell. Outgrowth is a period of active biosynthetic activity and is markedly inhibited by interference with the energy supply or by antibiotics which inhibit cell—wall, protein, or nucleic-acid synthesis.

If heat-activated spores are germinated under suitable conditions a high degree of synchrony is obtained. Such a synchronous population provides a model system for the study of the initiation of transcriptional and translational events occurring during differentiation. Since the dormant spore is devoid of functional mRNA the conversion of the spore to a vegetative cell during outgrowth requires transcription and de novo synthesis of gene products. The appearance of these gene products is ordered, determined by the time of transcription of particular genes. Ribosome synthesis starts early; vegetative cell-wall synthesis later, and DNA synthesis begins just before division of the cell. During a stepwise doubling of cell members for several generations, the initiation of certain enzymes occurs at a specific time during each division cycle, and results in a doubling of each enzyme during only a fraction of the total cycle.

Factors Influencing Microbial Growth

The rates of microbial growth and death are greatly influenced by a number of environmental parameters, some environmental conditions favoring rapid microbial reproduction and others not permitting any microbial growth. Conditions permitting the growth of one microorganism may preclude the growth of another. Not all microorganisms can grow under identical conditions. Each micrograms has a specific tolerance range for specific environmental parameters. Outside the range of environmental conditions under which a given microorganism can reproduce, it may either survive in a relatively dormant state or may lose viability; that is, it may lose the ability to reproduce and consequently die.

In both laboratory and natural situations some environmental parameter or interaction of environmental parameters controls the rate of growth or death of a given microbial species. In nature, where condition cannot be controlled and many species coexist, fluctuation environmental conditions favour population shifts because of the varying growth rates of individual microbial populations within the community of a given location. In the laboratory it is possible to adjust conditions to achieve optimal growth rates for a given microorganism. Similarly, in industrial fermentors conditions can be adjusted to optimize microbial growth rates, thereby maximizing the accumulation of desired microbial metabolic products. Many laboratory and industrial applications use pure cultures of microorganisms, facilitating the adjustment of the growth conditions so that they favour optimal growth of the particular microbial species.

Effects of Temperature on Microbial Growth Rates

Temperature is one of the most important of factors affecting both the rates of microbial growth and death. The minimum and maximum temperatures at which a microorganism can grow establish the temperature growth range for that microorganism. Most microorganisms are killed when exposed to temperatures above the maximal growth temperature. At very low temperatures, however, microbial metabolism ceases, preserving the microorganisms in a dormant state. Some strains of microorganisms in culture collections are maintained indefinitely by freezing them at very low temperatures, such as in liquid nitrogen; such preservation of reference type cultures is important because it provides necessary standards.

Within the growth range each microorganism has an optimal growth temperature at which the highest rate of reproduction occurs. The optimal growth temperature is defined by the maximal growth rate shortest generation time), not the maximal cell yield. Sometimes, greater cell or product yields are achieved at lower or higher temperatures. In the laboratory, incubators (controlled temperature chambers) are normally used to establish conditions that permit the growth of a microbial culture at temperatures favourig optimal growth rates. Different microorganisms have different optimal growth temperatures.

Some microorganisms grow best a low temperatures. Such organisms, known as psychrophiles, have optimal growth temperatures of under 20°C. Psychrophilic microorganisms have enzymes that are inactivated at even moderate temperatures, about 25°C. Some psychrophilic microorganisms are capable of growing below 0°C as long as liquid water is available. Psychrophilic microorganisms are commonly found in the world's oceans and are also capable of growing in a household refrigerator, where they are important agents of food spoilage.

Other microorganisms, known as mesophiles, have optimal growth temperatures in the middle temperature range between 20 and 50°C. Most of the bacteria grown introductory microbiology laboratory courses are mesophilic. Many mesophiles have an optimal temperature of about 37°C, which corresponds, to human body temperature. Many of the normal resident microorganisms of the human body, such as *E. coli* are mesophiles. Similarly, most human pathogenic microorganisms are mesophiles and thus are able to grow rapidly and establish an infection within the human body. Although some microorganisms are restricted to growth near the optimal growth temperature, others can grow over a wide range of temperatures. For example, psychrotrophic microorganisms have an optimal growth temperature between 20 and 50°C, but they are capable of growing at low temperatures (about 5°C). as inside a refrigerator.

Whereas mesophiles grow in the middle temperature range, thermophilic microorganisms grow at higher temperatures, often growing only above 40°C. Many thermophilic microorganisms have optimal growth temperatures of about 55-60°C. One finds thermophilic microorganisms in such exotic places as hot springs and effluents from laundromats. Recently bacteria have been found that can grow above 100°C. These bacteria live at great ocean depths, near thermal rifts, where the water is hot and still liquid because of the very high pressure.

Oxygen

Another factor that greatly influences microbial growth rates is the concentration of molecular oxygen. Microorganisms can be classified as aerobes, anaerobes, facultative anaerobes, or microaerophiles based on their oxygen requirements and tolerances. Aerobic microorganisms (aerobes) grow only when oxygen is available to support their respiratory metabolism, whereas anaerobic microorganisms (anaerobes), such as *Clostridium*, grow in the absence of molecular oxygen. Anaerobic microorganisms may carry out fermentation or anaerobic respiration to generate ATP. Some anaerobes have very high death rates in the presence of oxygen and such organisms are termed oxylabile anaerobes. Other anaerobic microorganisms, known as oxyduric anaerobes, although unable to grow, have low death rates in the presence of oxygen. In contrast to obligate anaerobes, which grow only in the absence of molecular oxygen, facultative anaerobes, such as *E. coli*, can grow with or without oxygen. As a rule, facultative anaerobes are capable of both fermentative and respiratory metabolism.

Although oxygen is required for the growth of many microorganisms, it can also be toxic. Some microorganisms, known as microaerophiles, grow only over a very narrow range of oxygen concentrations. Microaerophiles require oxygen but exhibit maximal growth rates at reduced oxygen concentrations because higher oxygen concentrations are toxic to these organisms. As a consequence of the toxicity of oxygen to microorganisms, even aerobic microorganisms generally posses enzyme systems for detoxifying various forms of oxygen. Oxygen can exist in a number of energetic states, some for which are more toxic than others. The common state of oxygen is the

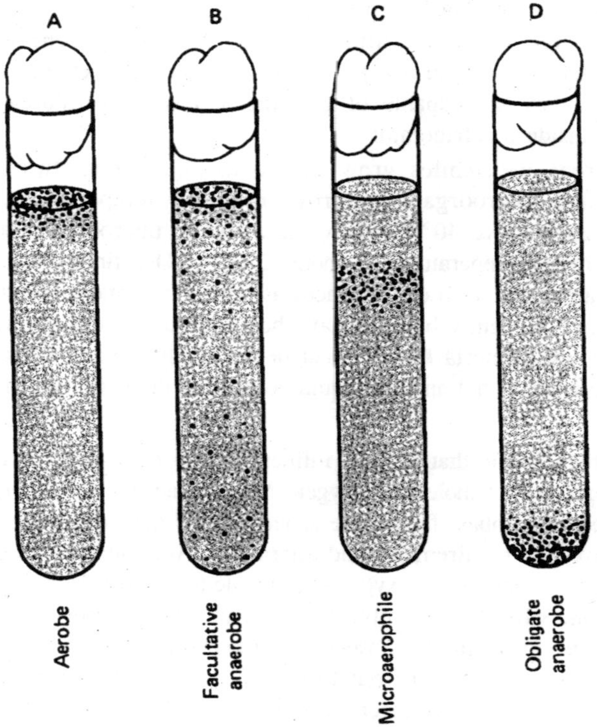

Fig. 6.3. These test tubes show the relationship of the growth of various types of microorganisms to the presence of oxygen.

triplet form in which two of the electrons in the valance shell are unpaired and spin in parallel directions. In singles oxygen, which has a higher energy level than the triplet form, these two electrons have antiparallel spins. Various enzymes are capable of catalyzing the conversion of triplet oxygen to singlet oxygen. The singlet oxygen form is chemically reactive and is extremely toxic to living organisms. Phospholipids can be oxidized by singlet oxygen, leading to a disruption of membrane function and the death of microorganisms. Peroxidase enzymes in saliva and phagocyte cells (blood cells involved in the defense mechanism of the human body against invading microorganisms) generate singlet, oxygen, accounting in part for the antibacterial activity of saliva and the ability of phagocytic blood cells to kill invading microorganisms.

Additionally, the conversion of oxygen to water, which occurs when oxygen serves as a terminal electron acceptor in respiration pathway, involves the formation of an intermediary form of oxygen known as the superoxide anion (O_2^-), in addition to forming singlet oxygen. Hydrogen peroxide and free hydroxyl radicals are also generated as intermediates in this process. These are extremely reactive oxidative chemicals that can irreversibly denature many biochemical. Hydrogen peroxide is frequently utilized as an antiseptic because it is toxic and kills microorganisms. Microorganisms generally contain enzymes that protect them against the toxicity of these forms of oxygen. Both catalase and peroxidase enzymes are involved in the destruction of hydrogen peroxide. Catalase converts hydrogen peroxide to water and triplet state oxygen. Microbial production of catalase can be demonstrated by adding a loopful of a microbial culture to a 3 per cent solution of hydrogen peroxide. The evolution of gas bubbles, oxygen, is evidence for the action of the catalase enzymes. Peroxidase uses the coenzyme NADH to convert hydrogen peroxide to water. In addition to these enzymes, the superoxide radical is converted to hydrogen peroxide and water by the action of the enzyme superoxide dismutase.

Both obligate aerobes and facultative anaerobes usually produce both catalase and superoxide dismutase enzymes. These enzymes permit such microorganisms to use oxygen and continue growing without accumulating toxic forms of oxygen that would kill the organism. In contrast, obligate oxylabile anaerobes generally lack both catalase and superoxide dismutase enzymes. The inability of these organisms enzymatically to remove toxic forms of oxygen probably accounts for the fact that they are obligately anaerobic and sensitive to oxygen. The oxyduric anaerobes generally produce superoxide dismutase but not catalase, evidence that they can survive in the presence of oxygen.

Water Activity

All microorganisms require water for growth and reproduction. Water is essential for most biochemical reactions in living systems, and the availability of water has a marked influence on microbial growth rates.

Water activity is an index of the water that is actually available for utilization by microorganisms. In the atmosphere, availability of water is expressed as relative humidity (RH).RH=100 × A_w; thus a relative humidity of 90 percent corresponds to an A*w* of 0.90. Pure distilled water has a water activity (A_w) of 1.0. Adsorption and solution factors, however, can reduce the availability of water and thus lower the water activity. Water, for example, may be bound by a solute and hence be unavailable to the microorganisms. Adding high concentrations

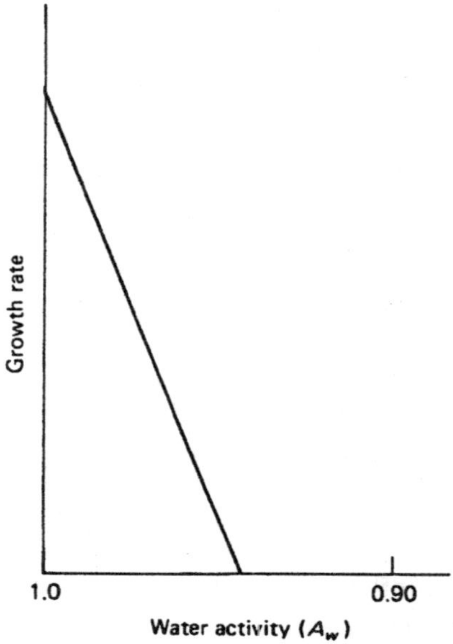

Fig. 6.4. The effect of water activity (A_w) on the growth rate of microorganisms.

of sugars, such as sucrose, to a solution lowers the availability of water. For example, maple syrup has a water activity of 0.9. Similarly, adding salt (NaCl) to a solution can lower the availability of water. A saturated solution of sodium chloride has an A_w of 0.8.

Most microorganisms requires a water activity above 0.9 for active metabolism. The relatively low availability of water in the atmosphere amounts for the inability of microorganisms to grow in the air. Microorganisms likewise are unable to grow on dry surfaces, except when there is a relatively high humidity. Microbial growth on surfaces is a problem in tropical zones, where the available water in the atmosphere can support microbial growth, permitting microorganisms to grow on clothing, tents, and numerous other surfaces where microbial growth normally does not occur in temperate regions.

Pressure

Osmotic Pressure

The solute concentration of a solution effects the osmotic pressure that is exerted across the cytoplasmic membrane of a microorganism. The cell-wall structures of bacteria and other microorganisms make them relatively resistant to changes in osmotic pressure, but extreme osmotic pressures can result in the death of microorganisms. In hypertonic solutions microorganisms may shrink and become desiccated; and in hypotonic solutions, the cell may burst. Organisms that can grow in solutions with high solute concentrations are called osmotolerant. These organisms are able to withstand high osmotic pressures and also grow at low water activities. Some microorganisms are actually osmophilic, requiring a high solute concentration for growth. For example, the fungus *Xeromyces* is an osmophile, and the optimum A*w* for *Xeromyces* is approximately 0.9.

Hydrostatic Pressure

In addition to osmotic pressure, hydrostatic pressure can influence microbial growth rates. Hydrostatic pressure refers to the pressure exerted by a water column as a result of the weight of the water column. Each 10 meters of water depth is equivalent to approximately 1 atmosphere pressure. Most microorganisms are relatively tolerant to the hydrostatic pressures in most natural systems but cannot tolerate the extremely high hydrostatic pressures that characterize deep ocean regions. High hydrostatic pressures of greater than 200 atmospheres generally inactivate enzymes and disrupt membrane transport processes. However, some microorganisms—referred to as barotolerant—can grow at high hydro-static pressures and there even appear to be some microorganisms—referred to as barophiles—that grow best at high hydrostatic pressures.

SALINITY

A special terminology is used to describe the salt tolerance and requirements of microorganisms. Some microorganisms, known as halophiles, specifically require sodium chloride for growth. Moderate halophiles, which include many marine bacteria, grow best at salt (NaCl) concentrations of about 3 percent NaCl. Extreme halophiles exhibit maximal growth rates in saturated brine solutions. These organisms grow quite well in salt concentrations of greater than 15 percent NaCl and can grow in places like salt lakes and pickle barrels. High salt concentrations normally disrupt membrane transport systems and denature proteins, and the extreme halophiles must possess physiological mechanisms for tolerating high salt concentrations. Fox example, the extreme halophilic bacterium. *Halobacterium*, possesses an unusual cytoplasmic membrane and many unusual enzymes that require a high salt concentration for activity.

Most microorganisms, however, do not possess these physiological adaptations and are not toluant of high salt concentrations. The degree

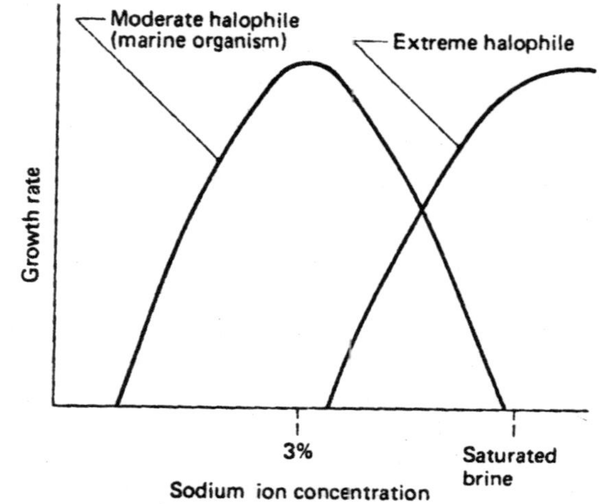

Fig. 6.5. The effect of salt concentration (sodium ion concentration) on the growth of microorganisms of varying salt tolerances.

of sensitivity to salt varies for different microbial species. Many bacteria will not grow at a salt concentration of 3 percent. Some strains of *Staphylococcus*, however, are salt tolerant and grow at concentrations greater than 10 percent NaCl. This physiological adaptation is *Staphylococcus* is important, as some members of this genus on skin surfaces where salt concentrations can be relatively high.

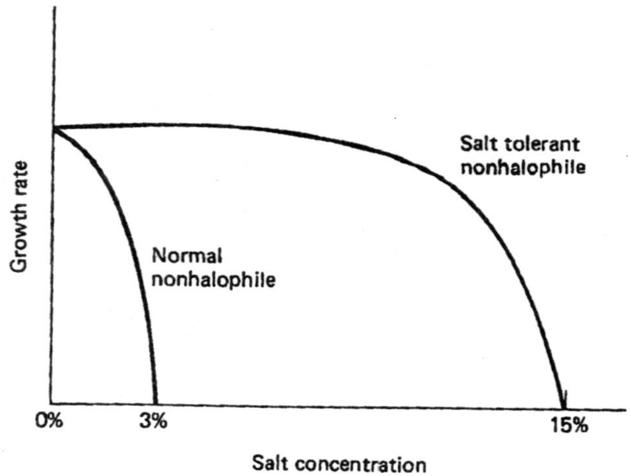

Fig. 6.6. Diagram showing ralative salt tolerance.

ACIDITY AND pH

The pH of a solution describes the hydrogen in concentration (H^+). The pH is equal to – log (H^+) or 1/log (H^+). A neutral solutions has a pH of 70 acidic solutions have pH values less than 7, and alkaline or basic solutions have pH values greater than 7.

Table 6.2. Table of pH Tolerances of Various Bacteria.

	pH		
Organism	*Minimum*	*Optimum*	*Maximum*
Thiobacillus thiooxidans	1.0	2.0—2.8	4.0—6.0
Lactobacillus acidophilus	4.0—4.6	5.8—6.6	6.8
Escherichia coli	4.4	6.0—7.0	9.0
Proteus vulgaris	4.4	6.0—7.0	8.4
Enterobacter aerogenes	4.4	6.0—7.0	9.0
Clostridium sporogenes	5.0—5.8	6.0—7.6	8.5—9.0
Pseudomonas aeruginosa	5.6	6.6—7.0	8.0
Erwinia cartovora	5.6	7.1	9.3
Nitrobacter spp.	6.6	7.6—8.6	10.0
Nitrosomonas spp.	7.0—7.6	8.0—8.8	9.4

Microorganisms vary in their pH tolerance ranges. Fungi generally exhibit a wider pH range growing well over a pH range of 5-9, compared in most bacteria, which grow well over a pH range of 6-9.

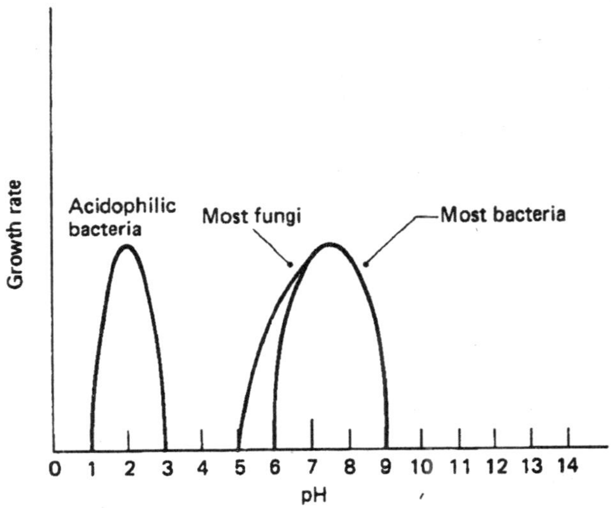

Fig. 6.7. The pH ranges for the growth of fungi and bacteria.

Similarly, some fungi grow well at lower pH values, as low as 0. Some other eukaryotic microorganisms, including protozoa an algae, are able to grow at low pH values; the lower limit for growth of some protozoa is approximately 2 and for some algae approximately 1. Although most bacteria are unable to grow at low pH, there are some exceptional cases.

Some bacteria tolerate pH values as low as 0.8. There are even some bacteria, called acidophiles, that are restricted to growth at low pH values. Some members of the genus *Thiobacillus* are acidophilic, and grow only at pH values near 2. Acidophilic bacteria possess physiological adaptations that permit enzymatic and membrane transport activities. The cytoplasmic membrane of an acidophilic bacterium breaks down and cannot function at neutral pH values.

Light Radiation

Although exposure to high intensities of light can be lethal to some microorganisms, photosynthetic microorganisms require light in the visible spectrum to carry out photosynthesis. The rate of photosynthesis is a function of light intensity. At some light intensities, rates of photosynthesis reach a maximum, and although light intensities above this level do not result in further increases in the rates of photosynthesis, light intensities below the optimal level result in lower rates of photosynthesis.

The wavelength of light also has a marked effect on the rates of photosynthesis. Different photosynthetic microorganisms are capable of using light of different wavelengths. For example, anaerobic photosynthetic bacteria use light of longer wavelengths than eukaryotic algae are capable of using. Many photosynthetic microorganisms have accessory pigments that enable them to use light of wave-lengths other than the absorption wavelength for the primary photosynthetic pigments. The distribution of photosynthetic microorganisms in nature reflects the variations in the ability to use light of different wavelengths and the differential penetration of different colours of light into aquatic habitats.

Exposure to visible light can also cause the death of microorganisms. Visible light may be absorbed by various biochemicals of microbial cells, resulting in changes in the energetic states of those biochemicals. In some cases this can lead to interference with metabolic activities or damage to cellular structures. Additionally, exposure to visible light can lead to the formation of singlet oxygen. As indicated earlier in this chapter, single oxygen is chemically reactive and can result in the death of a microorganisms. Some microorganisms produce pigments that protect them against the lethal effects of exposure to light. For example, yellow, orange, or red carotenoid pigments interfere with the formation and action of singlet oxygen, preventing its lethal action. Microorganisms possessing carotenoid pigments can tolerate much higher levels of exposure to sunlight than nonpigmented microorganisms. Pigmented microorganisms are often found on surfaces that are exposed to direct sunlight, such as on leaves of trees. Many viable microorganisms found in the air produce coloured pigments. Likewise, some microbial spores that are principally used for aerial dispersal are pigmented.

7

CULTURE OF BACTERIA

Bacteria derive their nutrition from the substrate in which the grow but food requirements of pathogenic bacteria vary widely from organism o organism. Some pathogens can easily grow under wider ranges of nutrition and environments whereas others, which are strictly parasitic, can grow under limited nutritional and environmental conditions and require growth factors, optimum temperature and adjustment of pH of the media. It is, however, not possible to supply the exact counterpart of the requirements of food of bacteria *in vitro* as the pathogens would derive them by their growth in the tissues. But it is also a fact that these organisms have a high degree of adaptability and can grow fairly well in artificial medial. Media have been devised from the time of Koch and Pasteur and, by trial and error method, suitable media have been evolved for almost all organisms. The question of growing viruses in artificial media is difficult as they are strictly parasitic, pass a "borrowed life" and prefer to grow in living cells.

Media to Supply Food and Energy

Bacteria require carbon, nitrogen, oxygen, mineral salts, etc., for their growth and multiplication.

Carbon

It is one of the most important constituents of the bacterial cell. Being devoid of chlorophyll, it is rather impossible for the pathogenic bacteria to utilize CO_2 from air. They, therefore, depend on organic carbon derived by breaking down carbohydrate, generally supplied in the form of glucose and other sugars. It is needless to say that the growth is very much accelerated by the optimal supply of glucose in the media.

Nitrogen

Bacterial body consists of a large percentage of protein and for the synthesis of protein, supply of nitrogen is essential. Inorganic nitrogen like ammonia is utilized by autotrophes or saprophytes, but for the growth of pathogenic organisms supply of organic nitrogen is necessary. These are derived from meat, casein or other protein substances. Bacteria do not utilize the protein directly as their food, but before utilization of the protein by bacteria, it is made soluble so that they can derive their nutrition from soluble protein like peptone.

Peptone is a crude, soluble product of organic nitrogen and consists of a mixture of protease, peptone, polypeptides and amino-acids. Being soluble in water, it can be easily incorporated in media and is not coagulated by heat when subsequently sterilized. It also supplies amino acids in an assimilable form for the synthesis of bacterial protein. As an alternative method, instead of incorporating prepared peptone, media may be prepared by fresh digestion of meat or casein rich in organic nitrogen by the action of trypsin or pepsin to form simpler compounds. The medium prepared by this method is known as *digest broth*. Hartley's broth is prepared by this method by digesting meat with trypsin.

On the other hand, certain organisms prefer to grow in presence of coagulated native proteins or eggs albumen and such organisms grow I media like Loeffler's medium, Dorset's egg medium, as in the case of *C. diptheriae*, or *Mycotuberculosis*, and so on. Other substances that might be supplied to the medium and derived from fresh body fluid, namely, blood, serum, hydrocele or ascitic fluid. The ideal way of utilizing these is without heating. Heating destroys the accessory growth factors, which are often heat-labile. So in these cases, unnecessary heating should be avoided and if necessary, sterilization by a process of filtration may be utilized to make the fluid bacteria-free, before it is added aseptically to the medium.

Meat extract

This is used in the medium, either as meat infusion or as LabLemco, a commercial product. The value of meat extract has been fully recognized and when they are incorporated into the media, they show better growth than in those without it. The exact nature of the watery extract of meat in the role of nutrition of bacteria has not been fully determined, but it consists mainly of mineral salts, small amounts of sugar, and some soluble nitrogenous muscle extracts containing amino acids and growth factors.

Mineral salts

Inorganic substances like chlorides, phosphates and sulphates play important roles in nutrient media, namely, (a) maintenance of suitable osmotic pressure of the cell and acid base balance by acting as buffers, (b) to function as enzyme or coenzyme; iron, magnesium, sulphur, zinc, manganese, etc., fall in this category. Iron forms the basis of catalase (haematin enzyme), peroxidase, cytochrome (haematin protein) and cytochrome oxidase. Pappenheimer and Johnson (1936) could produce highly potent toxin or synthetic media by limiting the supply of iron. Too much iron, however, interferes with the toxin production of bacteria. Magnesium activates phosphatase; copper in essential for the production of polyphenol-oxidase, and zinc for carbonic anhydrase. All these metallic ions, therefore, play essential parts in the production of enzymic structures. Calcium helps in the action of lecithinase.

Growth factors

Some highly pathogenic organisms require growth factors in the media for their growth and multiplication, e.g., *H. influenzae* requires X and V factors. The X factor is a heat stable substance which resists autoclaving at 120°C. It consists of haematin and is likely to help synthesis of catalase, which is essential for the growth of the organism is presence of oxygen. The 'V' factor is coenzyme I, heat labile and present in all living cells. It plays an important part in bacteria respiration.

Use of para-aminobenzoic acid in media

Blood culture or culture of pathological materials from cases treated with sulphonamides, may not show any growth of the organisms due to the presence of the drug in blood in a variable concentration. Thus they prevent the growth of bacteria. Sulphonamide competes with *p*-aminobenzoic acid required by the organism and addition of one molecule of *p*-aminobenzoic acid (PABA) counteracts the bacteriostatic action of 26,000 molecules of Sulphonamide. For routine culture, 5 mg of PABA is used per 100 ml of broth. It is heat stable and withstands autoclaving very well. As sulphonamides are extensively used in practice nowadays, it is better to use PABA in this proportion as a routine for culture of blood, cerebrospinal fluid and other pathological materials.

Synthetic salt media

At one time it was thought that for the growth of organisms and production of toxin, peptone-like substances produced by enzymic digestion of meat, were essential; but the work of Mueller and Miller (1946) on the nutritional requirements of *C. diphtheriae* in synthetic media led to a great advancement of our knowledge of media. The authors produced a good growth and a potent toxin in synthetic media containing known amino-acids, inorganic salts and maltose. In this medium, they also worked out growth factors for *C. diphtheriae* like pimelic acid, nicotinic acid, β-alanine and iron.

Growth factors of bacteria have bee studies by adding substances in definite quantities in basal synthetic media. By quantitative addition, the exact requirement of amino-acids and growth factors for bacterial growth, toxin production, and enzymic activity can be estimated in proportion to their concentration in the medium.

Special Culture Conditions

Synchronous growth

Even during the log phase cells are found at each stage along the division cycle. In order to study the division process and the time of production of specific proteins during the cell cycle, it is desirable to obtain synchronous cultures of cells. These are cultures in which all cells are simultaneously at approximately the same stage of the division process. In such a culture, the number of cells will remain constant and then double in a step-wise fashion within a short period of time. It is not possible to maintain synchronous growth over a long period, but several cycles of synchronous doublings usually can be obtained. It should be noted, however, that during synchronous growth the bacterial mass increases roughly exponentially throughout the step-wise growth process. A variety of methods have been employed for obtaining synchronous cultures, such as chilling the culture for a short period or limiting a required nutrient. In another method, cells are absorbed onto the surface of a filter followed by backwashing with growth medium, resulting in the elution of only newly divided cells. Those collected over a short period of time are in a very similar stage of the cell cycle and will produce several cycles of synchronous growth. A cell population can also be separated centrifugally according to cell size on a sucrose gradient, providing cells at defined states of division; these also will exhibit synchronous growth for several generations.

Continuous growth

For most studies of the physiology of bacteria, it is desirable to maintain the cells constantly in the logarithmic phase. In the usual batch cultures, however, this is not possible because the medium and the cell population density are continually changing. A widely used continuous culture device used to maintain constant cell number is called a chemostat. This instrument provides for the continuous addition of fresh medium containing an excess of all of the required nutrients except one. The rate of addition of this limiting nutrient determines the growth rate and thereby the cell density.

Effect of growth rate on cell size and composition

Using chemostats and continuous growth techniques, as well as batch cultures in media of different compositions, it has been possible to define the effect of growth rate on the size and composition of some bacterial cells. Faster growing cells are larger, contain more DNA per cell, and more RNA per unit of DNA. This raises an interesting point. As an medium becomes richer and cells grow faster, the rate of DNA replication must increase proportionally. However, there is a maximal rate at which the chromosome can be replicated, which in the case of *E. coli* is about 30 minutes. Thus, to grow faster than this, the cells must possess multiple copies of their chromosome and must initiate a new round of replication before the previous round has been finished. As a result, fast-growing cells partition several chromosomal copies into each progeny, some of which are still in the

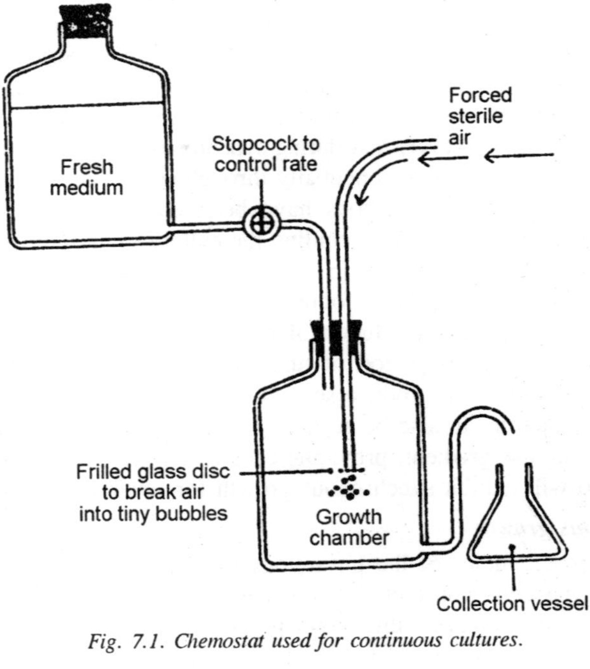

Fig. 7.1. Chemostat used for continuous cultures.

process of replication. For this reason, the control of cell division must be responsive not to the start of DNA replication but to the completion of a chromosome. The increased amount of RNA per nucleus reflects the increased number of ribosomes necessary to produce protein at the faster rate required in richer media.

Bacterial Nutrition

In this section, the nutrients that must be provided in a medium to obtain bacterial growth will be considered.

Carbon Source

For al the organisms of medical importance, although not for all bacteria, an organic source of carbon must be provided. This usually is both the source of carbon for most biosynthetic processes and the material whose metabolism yields most or all of the energy required for cell growth. With the possible exception of some plastics, there probably is no carbon-containing compound that cannot be used by some bacterium as a carbon source. Some bacteria can degrade hydrocarbons, such as asphalt or gasoline; others can digest wood. Most pathogens, however, are limited to metabolism of simple carbohydrates or amino acids.

Nitrogen Source

Since proteins and nucleic acids and many other cell constituents contain nitrogen, substantial amounts, of nitrogen must be provided for cell growth. A few free-living organisms can convert the nitrogen gas in the atmosphere into ammonia, which is then incorporated into cellular constituents. Many other bacteria can use inorganic sources of nitrogen, such as ammonium ions or nitrate. Other, less versatile bacteria require organic nitrogen, often in the form of glutamine or glutamate, asparagine, arginine, or peptide digests.

Inorganic Ions

All organisms require phosphate, both as a component of lipids and nucleic acids and for the storage of energy and the metabolism of carbohydrates. Sulphur, usually in the form of sulphate, can supply the sulphur present in the amino acids cysteine and methionine. Other ions are required to serve as cofactors for numerous enzymes; the required additions to any medium include Mg^{++}, K^{+}, and, occasionally, Na^{+} and Ca^{++}. Many other ions are required in such minute amounts that contaminating amounts present in the water and other ingredients provides sufficient levels of these trace elements. These include Mn, Zn, B, Se, Mo and Cu.

Standardization of Media

Media are to be standardized for their nitrogen content and for *p*H, i.e., H-ion concentration. Nowadays, peptones of various makes are available ready-made and, therefore nitrogen estimation is not necessary when peptone of a particular make is used. But it is better to standardize digest broth by estimation of the nitrogen content. This is matter of technical detail and beyond the scope of this book.

Reaction of media

True acidity of any fluid depends on the number of dissociable H-ion or hydrogen present in the solution as is measured by concentration of H-ion of the fluid. The greater is the concentration of H-ions, the more acidic is the medium. The following facts could be gathered from the ionization of conductivity.

(i) In conductivity, water dissociation of H-ion and hydroxyl or OH-ions occur in an equal proportion and exactly balance one another. (ii) Greater part of water is undissociable and equilibrium exists between the H and OH ions in this water. (iii) At equilibrium, the product of concentration of H and OH ions is constant and this is known as *ionization* (*dissociation*) *constant* or *Kw*. From conductivity measurement, it has been found that in pure water the concentration of H-ions is 10^{-7}, i.e., in one liter of pure neutral conductivity water there is an yield of 1/10 grammer of H-ions and also an equivalent weight of OH-ions, But, when hydrochloric acid is added, to this water there is an accelerated dissociation of H-ions, depending on the amount of acid added. The H-ion concentration is accordingly increased whereas the OH-ion is proportionally reduced. When an alkali like sodium hydroxide is added to distilled water, concentration of sodium hydroxide added and, therefore, there is also a corresponding decrease of H-ions. In case of conductivity water concentration of H-ions × concentration of OH-ions = ionization constant $Kw = 10^{-14}$ at 22°C.

The acidity of a solution is, therefore, best expressed in terms of H-ion concentration. It is inconvenient to express as 10^{-7} or 10^{-8} and to simplify the expression, pH or the H-ion exponent scale is used. Therefore, the pH value of a liquid is defined as the logarithm of the reciprocal of H-ion concentration. In other words, *p*H of a liquid = $\log (1/[H^+])$. For neutral pure water where concentration of hydrogen is 10^{-7} gm. per liter, the *p*H = $\log (1/10^{-7})$ or 7. Thus in an acid solution with concentration of hydrogen ions as 10^{-6} per liter, the *p*H is $\log (1/10^{-6})$ and expressed as 6. Again in an alkaline solution with hydrogen ion concentration as 10^{-8} per liter, the *p*H is $\log (1/10^{-8})$ and expressed as 8. As the scale is a logarithmic one, 1 units unit means a tenfold change of H-ions. Thus a liquid at *p*H 6 is 10 times more acid and one at *p*H 8 is 10 times more alkaline than one at *p*H 7. Estimation of H-ion concentration can be done by the following methods:

p*H meter*

The use of pH mete with 'quinhydrone' or 'glass electrode' is meant for accurate works. It has the advantage that exact dissociation of H-ions is recorded with these instruments and estimation of *p*H of coloured solution may be made without any difficulty. But it has disadvantage that the instrument is costly and all laboratories are not equipped with it.

Colorimetric method

This is done by comparison of colours in a comparator and for this an ordinary wooden block containing holes for test tubes in used. Lovibond comparator may be used conveniently for this purpose.

Media may be adjusted to a suitable *p*H with the help of known indicators, which give definite ranges of colour changes at different *p*H. If alkaline solution is gradually added to an acid solution containing phenol red, the colour becomes pink at *p*H 6.8 and purplish pink until the final range (*p*H 8.4) is reached. After this, there is no more change by addition of alkaline solution. Thus change of *p*H within this range can be estimated by comparing the colour with a standard solution. Estimation of *p*H by such method suits very well for the adjustment of *p*H of ordinary media ranging from *p*H 7.2 to *p*H 7.6. But media are sometimes coloured and in this case, it is difficult to judge the end point. Moreover buffering often interferes with the change of *p*H, when addition of large amount of acid or alkali may not alter the *p*H. In such a case, alteration of *p*H is very slow and the change of colour is negligible. The list of indicators that are generally used is given in Table 7.1 with variations of colour ranges at different *p*H.

Determination of pH of Media

1. *Equipments*. A set of standard tubes containing buffer solutions of known pH to which the indicator has been added. The tubes are of standard bore and contain phosphate buffers, ranging from pH 6.6—8 at intervals of 0.2. The tubes are used to match various colour changes in the range of the indicator.
2. *A comparator rack*. It may be planned in a wooden block with six holes to contain six tubes.
3. Four tubes of the same bore and the same glass as those containing the standard solutions are used to contain media, etc.

Table 7.1. Change of colour with different indicators at different *p*H

Name	*Range of pH*	*Change of colour*
B.D.H. Universal	3.0–11.0	Red–orange–yellow–green–blue–reddish-violet
Thymol blue (acid range)	1.2–2.8	Red to yellow
Bromophenol blue	2.8–4.6	Yellow to violet
Bromocresol green	3.6–5.2	Yellow to blue
Methyl red	4.4–6.2	Red to yellow
Bromocresol purple	5.2–6.8	Yellow to violet
Bromothymol blue	6.0–7.6	Yellow to blue
Phenol red	6.8–8.4	Yellow to purple pink
Thymol blue (alkaline)	8.0—9.6	Yellow to blue
Phenolphthalein	8.3—10.0	Pink to red
Litmus	4.5—8.3	Red to blue

4. *Indicator solution* used for this is 0.01 per cent of phenol red. This is available as a ready made solution or may be prepared as follows:

   Take 0.1 gm of phenol red and to this accurately add 10 ml of N/10 NaOH and 20 ml of distilled water. Dissolve it in a water-bath by gentle heat. Transfer the whole in a 500 ml volumetric flask washing out all the indicator into the flask. Now add accurately 10 ml of N/10 HCl and fill the flask with distilled water up to the mark. Then 0.1 per cent solution of phenol red is made by diluting this stock solution with an equal volume of distilled water.
5. *Microburette*. Each division of it should be 0.01 ml preferably.
6. N/20 NaOH solution.

Method

Take 5 ml of the medium in each of the standard test tubes, namely, 1, 3, and 5 and put 5 ml of distilled water to tube No. 4. To tube No. 3, add 0.5 ml of phenol read solution (0.01 per cent). Put standard pH tubes in 2 and 6. Run from the burette N/20 NaOH solution to tube No. 3, mix and continue to run the solution from burette till the colour has matched with the standard pH tube. It is easier to match the solution between the two standard colours than to match it with a given standard. When the colour has matched, take the reading of the burette. By this arrangement, tubes No. 2 and 6 contain standard tubes of the higher and lower pH values with a difference of 0.2 pH on either side of tube No.3 for comparison to obtain the desired pH. Tubes No. 1 and 5 containing media are superimposed on 2 and 6 respectively so that the effect of colour of the medium is duly considered in each case. The tube No. 4 (containing distilled water) is superimposed on the tube No. 3 (containing the medium to be standardized).

Calculation

Let the number of ml of N/20 NaOH run rom the burette to 5 ml of the medium to match the colour at pH 7.6 be *X*. That is 5 ml of the medium required *X* ml of N/20 NaOH. Then 1000 ml of the solution will require (1000 × *X*ml/5 × 20) N or 10 X ml of N/1 NaOH, which should be added to the medium (per litre) to obtain the desired pH.

Lovibond comparator

A Lovibond comparator may be conveniently used for the purpose of colour matching, Phenol red disk is made of standard coloured glasses of various shades. For each indicator a disk is available to match the colour. After matching the colour, the pH of the solution is automatically noted in the disk. Of course, in matching the colour the natural tint of the medium should be considered by superimposing a tube containing the medium on the tinted glass.

Clearing of media

The commonest method of clearing an agar medium is by putting the white of an egg in the medium. By this, finer particles present in the medium in suspension are avoided by coagulation of protein and filtration removes the particles entangled in the coagulum. White of two eggs is beaten with water and added to a litre of melted nutrient agar when it is cooled to 50°C to 55°C. The egg white is thoroughly mixed with the medium by shaking. The flask is then steamed in Arnold's staemer for half an hour. The medium is vigorously shaken to break the coagulum that has formed by steaming. It is steamed again for 15 minutes and filtered.

Filtration of media

(a) *Filtration through cotton-wool.* A suitably sized square piece of absorbent cotton is cut and split horizontally into two layers giving two squares of small size. One of these is put at the cone of a large funnel and another at right angle to the former. The cotton-wool is set by running distilled water and thus the filter-bed is formed. The medium is then slowly poured and filtered through cotton-wool. The first portion of the filtrate may not be clear and may require filtration for the second time: but the filter soon becomes efficient as the coagulum settles down, yielding a clear filtrate. In case of gelatin agar; filtration is best done in a warm chamber by placing the funnel in a warm double-jacketed chamber through which steam of hot water is passed.

(b) *Filtration through paper.* It is done with ribbed filter paper for filtering agar media. Whatman No. 1 ribbed filter paper may be used for broth or salt solutions, but quicker method is to use this filter paper in a Buchner funnel utilizing vacuum for filtration. The filtration of the solution or media is carried out under suction and the filtrate is collected in the filter flask. Water suction pump is quite sufficient for this purpose.

Distribution of media

After the filtration, media are distributed in test tubes, flasks and screw capped bottles in suitable quantities. Distribution is conveniently done from a funnel through a jet connected with a rubber tube. The flow of the medium is controlled with a pinch-cock. Approximate quantities can be distributed into same calibre test tubes by measuring the length of the tube filled with the known quantity of the medium. But in case of accurate measurement, it can be conveniently done, running the medium from a burette. For agar slopes, take 6 to 8 ml quantities in test tubes of ¾'' × 6" to give a good slope up to the 2/3rd of the tube. Slope may be prepared in a small screw-capped Bijou bottle and is specially useful for those media, which suffer from the effect of evaporation, e.g., egg media, Loeffler's serum medium, Lowenstein-Jensen's medium.

Preparation of Cultural Media

Liquid Media

Organic media in the form of nutrient broth constitutes the basal organic medium for preparation of all other media including solid media. The synthetic medium is, however, prepared by combination of synthetic salts which act as a basal medium. This is used for the study of growth factors for a particular bacterium.

Meat infusion (Bouillon)

500 gm of beef or ox heart freed from fat is minced and mixed with 1 litre of water, left overnight in the refrigerator and the fat from the surface is skimmed off next morning. The liquor is drained through a piece of lint, squeezing the residue. Collect in a flask and

boil for 15 minutes, when it becomes brown in colour on account of alteration of hemoglobin. It is then filtered through coarse filter paper. The volume is made up to 1000 ml with distilled water. After filtration, the product is light yellow in colour and possesses a sweetish odour. Meat infusion is distributed in flasks for storage and autoclaved.

Lab-Lemco

It is a concentrated meat extract and this can be used instead of meat infusion.

Nutrient broth

(a) To 1000 ml of meat extract, add 10 to 20 gm of peptone and 5 gm of sodium chloride. Peptone is dissolved by steaming. The pH is then adjusted to 7.8. It is now filtered, distributed and properly plugged and autoclaved.

(b) *Wright's Method*. In this method, peptone and sodium chloride are added to the added to the meat infusion or preferable veal infusion before filtering. Boil to dissolve the ingredients and filter. The filtrate is adjusted to pH 8.3 with N/1 NaOH, using phenolphthalein as indicator and phosphate is precipitated by boiling. It is then filtered: the pH is adjusted to 7.8 according to the method already described.

(c) *Lab-Lemco Broth*. Mix 10 gm of Lab-Lemco meat extract, 5 gm of sodium chloride and 10 gm of peptone in 1000 ml of distilled water; boil till the ingredients are dissolved. Then adjust the pH, filter and sterilize.

Broth is distributed in (i) test tubes 6" by ½" to ¾" in containing 5 to 8 ml; (ii) flasks of 250 ml capacity containing 100 ml; (iii) screw-capped bottles of 6 oz. capacity containing 50 ml of the broth. The medium can be stored also in bulk, in bottles or flasks, only taking care that evaporation does not take place and for this the containers may be placed in the refrigerator where only little evaporation occurs. The medium that is stored in bulk may be utilized for the preparation of nutrient agar, etc.

Digest Broth

In routine media, peptone is used as a source of digested protein but in the digest broth, meat is digested and converted into peptone and amino-acids by the action of trypsin or pepsin. Addition of peptone is not, therefore, necessary. Hartley's digest broth, prepared by trypsin digestion of meat is useful for the growth of streptococci and production, of diphtheria toxin.

Hartley's digest broth

In this, ox heart or meat is digested with pancreatic extract of Cole and Onslow, which is prepared as follows: Pancreatic extract—500 gm of fat-free fresh pancreas in mixed with 1500 ml of distilled water and 500 ml of alcohol in a stoppered bottle and mixed well by shaking. It is allowed to stand for 3 days at room temperature, shaking several times during the day. After 3 days, the mixture is strained and filtered through Chardin filter paper. The extract may be used fresh or it may be preserved for future work. In the latter case, and pure HCl to the extract in proportion of 1 per cent. This produces a cloudy precipitate, which settles down in a few days and the fluid is then filtered off or kept as it is, without filtration. The extract keeps for several months in stopered bottles, when kept in a cool place. HCl prevents deterioration of trypsin. Preparation—To 1500 gm of minced veal or beef free from fat, 2500 ml of tap water are added. Mix, heat in a steam sterilizer, until the temperature is raised to 80°C. Add 2500 ml of 0.8 per cent cold sodium carbonate solution. Cool to 45°C and 3 hours. The liquid is frequently stirred while in the process of digestion. After the digestion is complete, add 40 ml of pure HCl. Steam for half an hour and filter. Now adjust the reaction to alkaline side by NaOH solution to pH 8.4 when it is pink to phenolphthalein; steam for one hour to precipitate the phosphates. Filter and adjust the pH to 7.6. The digest broth so obtained is distributed in containers (tubes, flasks, etc.) and sterilized in an autoclaves Streptococci grow well in this medium. The medium is used specially for the production of diphtheria toxin.

Glucose broth

1 per cent glycerine is added to nutrient broth, distributed in test tubes or flasks which are plugged and autoclaved.

Bile broth

Add 0.5 per cent sodium taurocholate to nutrient broth and distribute in 50 ml and 100 ml quantities in 100 ml flasks and 8 oz. screw-capped phials respectively and autoclave.

Sugar media

For testing the biochemical reaction, sugar media are used. Some organisms produce acid only, but others produce both acid and gas in certain sugar, which is detected by the presence or gas in Durham's fermentation tube. Sugar solutions contain an indicator like neutral red or Andrade's indicator which becomes pink at a pH of 6.4. Different sugars are used for this purpose and are incorporated in the proportion of 0.5 to 1 per cent into 1 per cent peptone solution.

The pH is adjusted at 7.6. The indicator that is used is 0.25 per cent of 1 percent solution of neutral red or 1 per cent Andrade's indicator.

Sterilization

Sterile test tubes are used for putting the media and are tubes the properly plugged after putting the medium. The medium is sterilized by fractional sterilization method in Koch's sterilizer. It is steamed for half an hour for three successive days. It is steamed for half an hour for three successive days. Steaming drives away the air from Durham's tube.

Hiss's serum water

This medium is specially used, when the organism is a delicate one and the growth does not occur in ordinary peptone water. Therefore, further protein in the form of serum of sheep, horse or ox is added aseptically to sugar tubes (1 part of serum and 3 parts of sugar media) and media are sterilized.

Composition. Clear ox serum — 1 part
Sugar media — 3 parts
(for fermentation).

Steam for 15 minutes. This will destroy any diastatic ferment that may be present in the serum. The tubes do not show any coagulation if the medium is not acid.

Use. It is used for the culture of *C. diphtheriae*, *N. meningitidis*, *N. gonorrhoeae*, streptococcus, *D. pneumoniae*, etc., for sugar fermentation.

Litmus milk

In this, litmus powder is added to detect the reaction after fermentation of lactose, when it becomes red or may show clotting or peptonization of milk. Milk is steamed for 20 minutes, and put in the refrigerator overnight. Siphon it, add 2.5 per cent alcoholic solution of litmus; put in 3.5 ml amounts in sterile test tubes: steam for half an hour for 3 successive days. It is not advisable to preserve litmus milk in bulk as the colour fades away. Litmus milk can be prepared from skimmed milk powder by dissolving 150 gm of powder to a litre of water. Litmus milk prepared from such a milk can be autoclaved without any danger of discolouration.

Solid Media

These are obtained by addition of substances like agar-agar or gelatin to nutrient broth. Nutrient agar forms the basis of all solid media. Solid media are necessary for isolation of the organism in pure form and study of colonial morphology, preparation of saline suspension etc.

Agar-agar

This (popularly known as Chinese grass) is obtained from sea-weed and extracted by boiling it in water, allowed to gel in frozen state. The frozen material is then formed into shreds and placed in the sun for drying an bleaching. The material may be ground into small particles. Agar-agar is melted in a boiling water-bath (95°C) and again sets, when cooled to 40°C.

Nutrient agar.

Nutrient broth	1000 ml
Agar-agar	20 to 25 gm

Heat in a steam sterilizer for 1 hour, when it is fully dissolved. Clarify if necessary, with white of two eggs. Adjust the pH to 7.6 and filter through Whatman No. 1 filter paper. Distribute into tubes or flasks. Sterilize at 5 lb. Pressure for half an hour and prepare slopes or agar deep from the tubes.

Preparation of agar slopes

In order to prepare agar slopes, sterilized medium in tubes is allowed to set in a slanting position (8 ml in 6" × 3" tube). The slant should occupy a little more than half the length of the test tube. Frothing and air bubbles should be avoided. When the slopes have set, the tubes are stocked in a vertical position in a dark cool place.

Use. Slopes are used for stoke culture. *Semi-solid* or *Sloppy agar* (0.2 to 5% agar). Semi-solid nutrient agar may be distributed in 3 to 5 ml quantities as agar butts in narrow test tubes. (Butts are used for stab or shake culture).

Agar deeps are prepared from nutrient agar in 14 ml. quantities to pour into Petri dishes 10 cm diameter for preparing agar or blood agar plate.

Gelatin

It is an aqueous hydrolysate of protein obtained from bone, fascia and tendon. It is dried and then obtained in powered form. Commercial gelation is incorporated into broth for solidification of the medium.

Nutrient Gelatin (Koch)

It contains 15 to 18 per cent gelatin in broth. Gelatin is dissolved by heating in a steam sterilizer and the product is made just alkaline to phenolphthalein, as gelatin is acidic in reaction. The mixture is clarified with white of an egg and filtered in a warm funnel. The pH is adjusted to 7.6. and the medium is sterilized by fractional sterilization steaming for 20 minutes each day for three successive days. Autoclaving destroys the solidifying property of gelatin. Nutrient gelatin solidifies when cool but melts at about 24°C. Gelatin differs from agar in that it is liquefied by bacterial action. Study of liquefaction of gelatin should be done at a temperature of 18°C, or as an alternative, it may be incubated at 37°C, and the tubes are then placed in cold water to see whether it has undergone liquefaction or not.

Nutrient agar plates

For a Petri dish of 10 cm diameter 14 ml of the medium is sufficient. The agar deep is melted and poured into sterile Petri dish under aseptic condition.

Nutrient agar, blood or serum agar is allowed to set on the working table with the Petri dish lid slightly lifted, so as to allow the steam to escape. When set, the lid it put in position and the plates are stored in the refrigerator.

For use, the plates are dried in the incubator for about an hour to remove moisture. But they should be dried for 2 hours or so when spreaders (proteus group of organisms) are supposed to be present in the culture. Agar and blood agar plates are used for the culture of pathogenic organisms.

Enriched media

Delicate pathogenic organisms like *N. gonorrhoeae* and *N. meningitidis*, *H. influenzae, Bord. Pertussis*, etc., do not grow on ordinary nutrient agar unless it is enriched by adding blood, serum, hydrocele or ascitic fluid, etc. Enriched media are : blood agar, chocolate agar, serum agar, cystine agar, liver infusion agar (for brucella), Loeffler's medium for *C. diphtheriae*, Dorset's egg medium for *Myco. tuberculosis*.

Blood agar

Blood is usually added in the proportion of 7 to 10 per cent in the nutrient agar which has been melted and cooled at 50°C. Blood is mixed with melted agar avoiding frothing. Slopes and Petri dishes are then prepared aseptically.

In case, air bubble forms during pouring, it is immediately touched with a lot platinum loop to remove the air. Serum agar is prepared with serum instead of blood with same precautions.

Coagulated media

The nutrient media may be enriched by coagulating serum or blood (chocolate agar) into solid mass. The coagulation of serum or egg is effected by heating the tubes in a sloped position at 75°C to 85°C in an inspissator e.g., Loeffler's medium and Dorset's egg medium.

Differential and selective media

These are important media for isolation of intestinal pathogens avoiding the gross contamination of the commensals, which often outnumber the pathogens. The medium contains some inhibitory substances and indicators. The growth of the commercial organisms is inhibited and the pathogens get the opportunity to grow and multiply in preference to commensals. These are MacConkey's medium, desoxycholate citrate agar (DCA), Endo's medium, bismuth sulphite agar of Wilson and Blair, S.S. (Salmonella-Shigella) agar, etc. Tellurite medium is also a selective and differentiating medium for *C. diphtheriae*.

Enrichment media

In these, materials like stools are inoculated into liquid media for primary culture to enrich pathogens. It allows the growth of desired pathogenic organisms in preference to, or totally inhibiting, the contaminating ones. From this primary culture, subcultures are then made on differential and selective media. Selenite or Kauffman's tetrathionate broths are generally used as enrichment media for intestinal pathogens.

Storage of Stock Media

When a culture medium is prepared in bulk, a quantity our of it as required, may be stored in flasks or bottles in volumes which would be convenient for use. They are then autoclaved and kept ready in bulk. Broth, etc., may be distributed : (a) in 250 ml amount or 100 ml amount or, (b) in six ounces screw capped bottles in quantities of 100 ml. or, (c) is 5 ml and 8 ml quantities in 5/8" by 5" and ¾" by 6" tubes respectively as a stock for ready use.

Solid media

Agar may be stored in 100 ml quantities in flasks or screw capped bottles for preparing nutrient agar, blood agar, etc. For MacConky's medium, 100 ml is enough for 7 plates and it can be stored in 100 ml quantities in flasks.

Cold Storage

Use of a refrigerator in the laboratory is not only handy but essential. Formerly ice chests were used, but it had several drawbacks, the chief being that the ice was often exhausted particularly at the week-end and it did not maintain a low and uniform temperature. Refrigerators of various sizes may be used according to the articles to be stored. For a small laboratory, a domestic refrigerator may be used.

The principle of a refrigerator is that in these phreon gas is circulated by electricity, gas or kerosine oil. In commercial concerns, cold storage room with several thousand cubic feet are utilized with arrangement of circulation of ammonia or phreon gas. When a household refrigerator is used, it is necessary to defrost it to remove the ice at an interval of a fortnight or so.

After defrosting, the interior of the refrigerator is wiped dry with an absorbent cloth and the refrigerator is started again to store the articles. In addition to storage of sera, vaccines, bacterial suspensions and antibiotics, media, particularly blood agar and other delicate media, are stored without any deterioration or contamination. Evaporation of media is greatly prevented by storing them in the refrigerator.

Making of Culture

All media for culture should be fresh at the time of their use and slopes should have water of condensation at the bottom. Dry media without water of condensation are unsuitable for culture of delicate organisms. Media, therefore, should be preserved in a cool-place to avoid evaporation of moisture and extraneous contamination, dust, etc. The tubes are kept erect in a vertical position so that the water of condensation may not make the slope wet. In case of liquid media, soaking of the cotton-wool plug is avoided by keeping the tube erect.

Inoculation of Media

The process of inseminating bacteria in culture media is called *inoculation*. It may be done from pathological material or from another culture. The former is known as primary culture and the latter as *subculture*. Primary cultures are meant for isolation of pathogenic organism, but subcultures are made for further study of the isolated organism. A platinum loop is used for inoculating media.

Platinum loop

This consists of a platinum wire of 23 standard wire gauge (S.W.G.) thickness and about 3" long held on a metal holder or fused into a glass rod, which can be sterilized in naked flame. The free end is bent in the form of a loop of about 2 mm as internal diameter. The loop should be closed completely to hold a drop of liquid. It should be flat without any projection on the side, otherwise it may scratch or dig the medium. The platinum wire has the advantage that it cools very quickly after sterilization in a flame. A nichrome wire may be used and this is not as costly as the platinum wire; but the former has the disadvantage that it cools down slowly.

Use. The loop is heated in naked flame and the adjacent portion of the holder is similarly heated for sterilization. It is allowed to cool before it is charged by taking the material for inoculation. The inoculation is usually done on the slope by a process of stroke culture.

Spud

This is a straight needle bent at an angle and the tip is flattened with a hammer. This is used for scraping medium to pick up sticky colony, e.g., Actinomyces.

Straight needle

It is a wire similar to the loop held in a holder but the wire is straight in this case. It is sterilized like a loop.

Use. The bacterial colony can be exactly touched by the tip of the straight needle without touching medium as required in the case of inoculation of Koser's medium. It is generally used for a stab or shake culture.

Capillary pipette with rubber teat

A Pasteur pipette is used for this purpose. For use of a Pasteur pipette, a rubber teat is fitted to the plugged end an at the time of use, the capillary end is nicked with a glass file, snaped off and sterilized in the flame. It is used for inoculating pathological fluid like urine, cerebrospinal fluid, and exudate.

Inoculation

Inoculation of a medium should be done under aseptic precautions and the medium should be handled in such a way that extraneous contamination is avoided. The worker should, therefore, acquire the technique of handling the medium aseptically. The technique varies in primary and subcultures.

Isolation of Bacteria (Primary Culture)

Pathological materials or, materials like water, milk, etc., often contain pathogenic organisms as contaminants. Therefore, isolation of the pathogenic organisms in pure culture is necessary. Before the time of Robert Koch, there was no satisfactory method for the isolation of bacteria. Robert Koch utilized solid gelatin media for the purpose of isolation of microorganisms and since that time methods of isolation have developed.

Poured plate method

This method is chiefly used for making counts of viable bacteria in water. In this method, a serial dilution of the material is done and known amount of material is mixed with melted agar a temperature of 45°C to 50°C. The contents of the tube are mixed by rolling between the palms and poured into a Petri dish. The plate is then allowed to set. It is then incubated. Isolated discrete colonies appear in the medium, when the material is properly diluted.

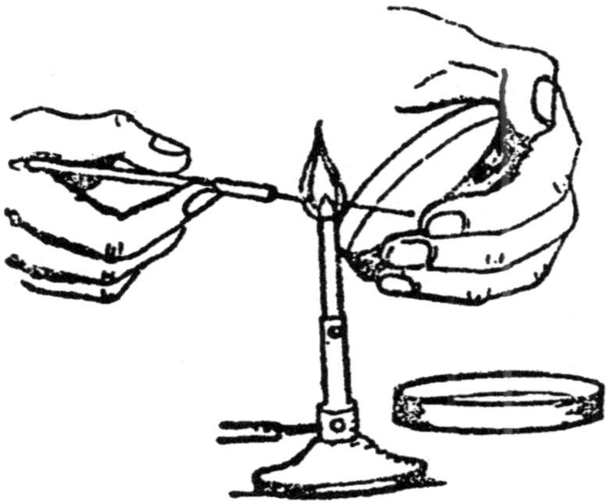

Fig. 7.2. Inoculation of an agar plate.

Shake culture

An agar deep is melted and then cooled down to 50°C, at which the organism will not be killed. The pathological material is inoculated into the melted agar and it is then rolled between the palms of the hand to be mixed with the agar. The tube is then allowed to set in the erect position and incubated. The colonies develop at the depth of the medium, wherefrom subcultures are made after breaking or cutting the tube. This method of culture is useful for anaerobic and microaerophilic organisms.

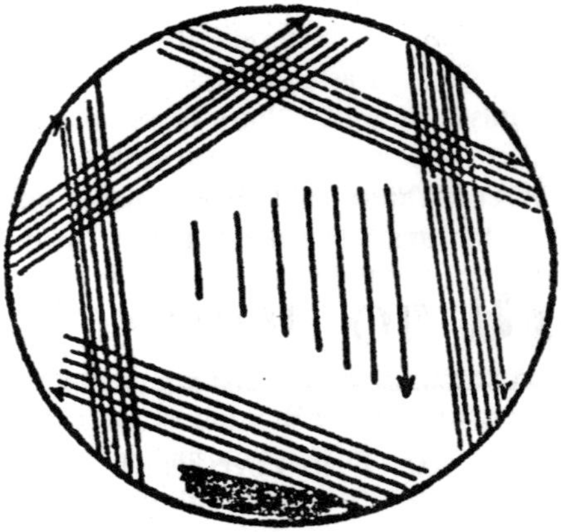

Fig. 7.3. Isolation by streaked plate met.

Streaked plate method

For this agar plates are used. A loopful of the material like faces, pus, sputum, etc., is taken. The material is rubbed in one segment of the plate. Stroke-cultures are then made. By this method, the number of organisms gradually becomes less and less and finally isolated single organisms are spread on the surface of the medium forming well-separated colonies on the agar plate. This is a routine procedure in the laboratory. Another way by streak method is shown.

Isolation by spreading

For this, a Petri dish of 6" diameter is selected and dried in the incubator for at least one hour. The plate is then divide on the glass side into 6 sectors with a glass pencil. The material is taken in a sterile Pasteur pipette and a drop or two from this put in one sector. The pipette is sealed at the tip and bent at a right angle so that it could be used as a spreader. Allow it to cool and ensure cooling to touching a part of medium. Spread out the material in the first sector. The spreader is then lifted and spread on the second sector which is similarly inoculated and so on to the third and subsequent sectors. The number of organisms are very scantly and isolated colonies develop after incubation.

Isolation by heating and subsequent plating

This is specially applicable to isolation of spore-bearing anaerobes. The vegetative form of the organism is killed by heating at 70°C for 10 minutes and the spores which survive this heating, later grow in the media or they may be plated for the growth of individual colonies.

Use of selective media

MacConkey's medium, desoxycholate citrate agar, Wilson and Blair's medium, etc., are used for the growth of intestinal group of organisms. Dieudonne's medium for *V. cholerae* and tellurite medium for *C. diphtheriae* have been used for isolation of organisms, as these media are differentiating and selective and allow growth of a particular organism suppressing others. Thus these media help in the isolation.

Addition of antibiotics

A few drops of penicillin are used in the cough plates (12 units if aqueous solution for a 10 cm plate) for isolation of *Bord. Pertussis.* Penicillin inhibits the growth of secondary organisms without affecting the growth of *Bord. Pertussis.*

Adding suitable chemicals to agar

Spreading of proteus group of organisms is prevented by adding sodium azide, aniline dyes, alcohol, or other antiseptics. Ten per cent salt broth may be used for isolation of staphylococci.

Animal inoculation

This method is sometimes used by inoculating sputum in mice for isolation of *D. pneumoniae.* Sputum is similarly inoculated into guineapigs for isolation of *Myco. Tuberculosis.*

Subculture

Inoculations are made under sterile and aseptic conditions. The loop is sterilized in the flame each time before an inoculation is made. The plug of the tube is burnt in the flame before opening the tubes, the mouth being towards the flame. For subculture from stock culture, the plugs of the test tubes are slightly loosened by the right hand with a twisting movement in order to break any possible adhesion of the plug with the wall of the tube. The culture tube and the tube to be inoculated are held at the bottom in a slanting position between the thumb and the other fingers of the left hand with the slope surface of the medium towards the worker. The mouths of the tubes held near the flame.

The tube containing the growth should be on the left and the tube to be inoculated on the right. The platinum wire is now held with the right hand, as if holding a pen and the wire including the adjacent part of the handle is burnt in the naked flame of a Bunsen burner. In inoculating from one tube to another, the cotton-wool plug of the tube from which inoculation is to taken, is removed by holding the plug in the web between the middle and ring fingers of the right hand. The

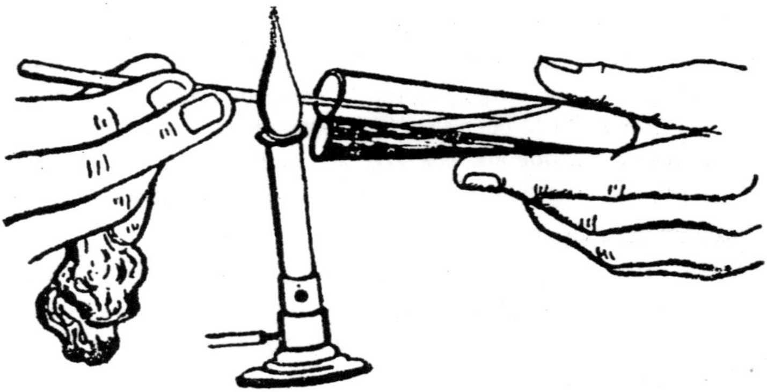

Fig. 7.4. Method of subculture on agar slope.

mouth of the tube is flamed and the platinum loop is immediately introduced into the culture tube, where it is held for a few seconds to allow it to cool or the loop may be cooled by touching a part of the medium, which is free from growth. This is very important as otherwise hot wire, touching a colony may kill the bacteria. The colony is touched or scraped with the sterile, cold loop and the loop is thus charged with the inoculum without disturbing the agar. The charged loop is then withdrawn and introduced into the agar tube after removing the plug with the crooked little finger of the right hand and flaming the mouth of the tube.

Now stroke culture is made on the agar slope with gentle parallel strokes of the platinum loop containing inoculum. The plugs are then replaced after gentle heating of the mouth. Care must be taken not to put the plug on the bench during the seeding of the tube, as this will always produce contamination. The loop is now withdrawn and sterilized by flaming. The nature of inoculum and date is then written. The inoculation is better done in an inoculation chamber, specially designed for it. This is necessary because draft of air or movement of the people might disturb the atmosphere acid, in such cases, there is every chance of aerial contamination.

Stab culture

In this case, the manoeuvre is the same as in the case of stroke subculture but a wire is used instead of the loop, for the purpose of inoculation. The colony is touched with the straight wire which is introduced vertically and then plugged into the centre of the agar but, care being taken to remove the wire in the same line as that of introduction.

Inoculation of a fluid medium

For this, the loop is charged with the inoculum, which is lightly touched on the side of the tube of flask at a level lower than the surface level of the broth so that the inoculation could be done on a moist surface, which comes in contact with broth, when the container is kept in an erect position.

Incubation

Most of the pathogenic organisms grow, best at or about body heat and for this cultures of pathological materials are kept at this temperature in a special chamber, known as the incubator, which maintains a constant temperature of 37°C. An incubator is essentially a double-jacket chamber made of copper sheets with an outer wooden or asbestos cover. The front part of the chamber is covered with a glass-door inside and a wooden door outside. The copper jacket contains water, which helps in maintaining a uniform distribution of temperature and moisture in the incubator. The interior of the chamber contains some adjustable shelves over which cultures are placed.

There are several types of incubators, heated by gas or electricity. The chief principle involved in a gas incubator is that the heating is done by gas burner fixed below. An expansile capsule automatically regulates the temperature of the chamber at 37°C with variation of temperature to fraction of a degree. Electrical incubators are simple in design. The chamber is made of copper sheet and the heating elements are put either inside (in some American models) or on the outer side of the copper chamber. The space between the inner copper plate and outer asbestos wall is packed with nonconducting substances. The temperature is controlled by a capsule. As the capsule is heated more than 37°C, the current is cut off but contact is re-established when the temperature falls below 37°C. There are models made of double copper jackets containing water. It has the advantage that heat is moist as water evaporates through holes in the inner copper plate at a level higher then water. If properly adjusted the variation of temperature in the incubator is only fraction of a degree.

Small rooms may conveniently be used as incubators with an automatic electrical arrangement for control of temperature and one such may be seen in the Pasteur Institute, Shillong.

Cold incubator

For cultivation of Leptospira, the culture tubes are incubated at 30°C. In the tropics, room temperatures is quite suitable for the cultivation of fungi. If an incubator is used, a cold incubator, with a temperature of 24°C to 25°C, is useful. For gelatin media, tubes are incubated at 18°C to 22°C as gelatin melts at a higher temperature. *Leishmania donovani* is cultivated in cold incubator (22°C to 28°C).

Anaerobic Culture

Strict anaerobes the those organisms which do not grow in presence of molecular oxygen. They prefer to grow in total exclusion of free oxygen. It has been found that, when these organisms grow in presence of atmospheric oxygen they develop hydrogen peroxide. This is due to the absence of catalase or cytochrome system in them. H_2O_2 is detrimental to the growth of that organism and prevents its multiplication. Anaerobic organisms are, therefore, cultured by:

Exclusion of atmospheric oxygen

(a) In this, the medium is a nutrient broth, containing 0.2 per cent agar (10 ml of 2 per cent agar and 90 ml of nutrient broth makes a sloppy agar). Air is removed from the medium just before use, by heating the medium in boiling water bath for 10 minutes. (b) contact with atmospheric oxygen can be cut off by putting a layer of liquid paraffin on the surface of the broth. This is used as additional measure in Robertson's cooked-meat medium.

Addition of reducing agent

These are glucose, sodium thioglycollate or ascorbic acid, cysteine, glutathione, etc. 1 to 2 per cent of glucose in sloppy agar medium may be used in Noguchi's tube (8" × 1/2"). It is used as a control culture for anaerobic test of vaccines, etc. Other reducing substances as 0.1 per cent sodium thioglycollate may be used, e.g., Brewer's medium. Recently 0.1 per cent ascorbic acid has been used as reducing agent. Before the use of sloppy agar with reducing agents and Robertson's cooked-meat medium, the tubes are kept in boiling water bath for 10 minutes to remove all traces of oxygen. These are then allowed to cool rapidly before inoculation. After incubation, the sloppy agar shows cracks if there is any gas-forming anaerobe.

Introducing catalase

This is supplied in the form of fresh animal tissue (rabbit's kidney) into the medium, e.g., Smith-Noguchi medium, or alkali treated bullock's heart muscle in Robertson's cooked-meat medium.

Displacement of oxygen

This is performed by passing a gas like hydrogen from a Kipp's apparatus into a Bulloch's jar for some time. Oxygen in this case is completely replaced by the gas. The remaining residual oxygen is absorbed by alkali pyrogallate solution.

Absorption of oxygen

For this, Buchner's tube is used. In this, a mixture of pyrogallic acid and NaOH is used. Alkaline pyrogallate solution absorbs a large quantity of free oxygen. Buchner (1881) introduced this principle. The apparatus consists of a stout-walled large test tube 8" × 2", with a well-fitted rubber stopper; the lower end of the tube is blown into a buld and, above the bulb there is a ledge, over which the inoculated culture tube rests. For use, some pyrogallic acid is put in the bulb and 25 per cent NaOH is then added to fill half of the bulb producing alkaline pyrogallate which absorbs oxygen. The inoculated tube is then quickly introduced into the Buchner's tube which is immediately closed tightly with the rubber stopper and finally sealed with melted paraffin. The apparatus is then incubated at 37°C in an incubator.

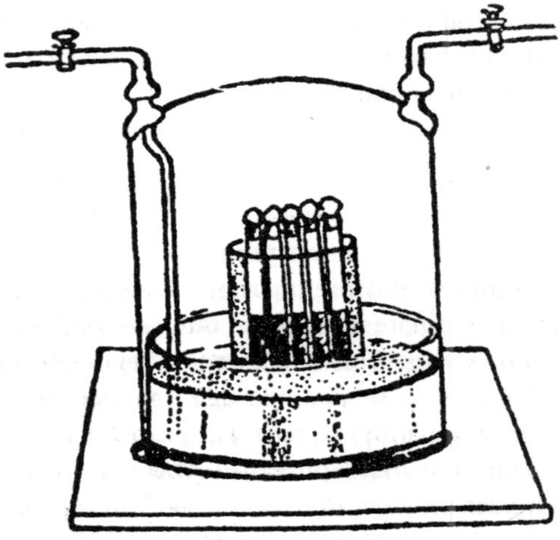

Fig. 7.5. Bulloch's jar.

Displacement of oxygen combined with combustion

The method is adopted in McIntosh and Fildes' Jar. The apparatus consists of stout thick glass or metal jar, 8" by 5" in diameters, with a 'Press on' metal lid which can be clamped tightly. The lid is furnished with two gas-tight needle-valves, one for the inlet and the other for the outlet of gases. A small glass or porcelain spool with spongy palladinized asbestos around it, suspended from the lid by means of two wire, are connected with two terminals fitted on the lids. A fine coil of resistance wire is wound round the palladinized asbestos and ends are connected with the wire supporting the spool, so that an electric current can be passed and spongy palladinized asbestos in heated to cause slow combustion of oxygen in presence of hydrogen to produce H_2O. The wire gauze round the spool acts on the principle of Davy's safety lamp and it prevents an explosion which might occur by a mixture of hydrogen and oxygen. The amount of electric current passed is reduced by 3 carbon filament lamps in parallel series (two 16 and one 8 c.p.). With this arrangement, it may be connected with ordinary domestic electric supply. Inoculated test tubes, Petri dishes, etc., for anaerobic culture are placed inside the jar and so also an indicator test tube, containing equal volumes of (a) 6 ml of N/10 NaOH diluted to 100 ml, (b) 6 gm of glucose diluted to 100 ml and (c) 3 ml of 0.5 per cent watery solution of methylene blue diluted to 100 ml. The mixture is heated till it is colourless and a few crystals of thymol are added as a preservative. The indicator tube is at once placed in the jar. It should remain colourless when the current is passed and the anaerobic condition is complete.

Technique

The lid is smeared with vaseline and clamped down into the jar tightly. A considerable saving of the time and hydrogen is effected, if the jar is connected with a suction pump to remove the bulk of air from the jar at the time of passing hydrogen. Hydrogen is passed from a cylinder containing compressed gas to the chamber with a negative pressure of 720 mm of Hg. This is obtained after suction for 15 to 20 minutes. Hydrogen gas is passed through the needle valves either directly from a Kipp's apparatus or preferably from a hydrogen cylinder into the jar, avoiding too much pressure. The best way to operate is to connect the cylinder through an ordinary football bladder, introduced between the jar and the cylinder, by means of a "T" tube. The gas is slowly turned on and the bladder in inflated. The gas is slowly turned on and the bladder is inflated.

The gas is then turned off. The tube of the bladder connected with the cylinder is closed by a screw clamp and hydrogen flows from the inflated bladder to the jar. The bladder is refilled with gas as the pressure becomes less. The connection with the jar through the series of resistance lamps is switched on so that, during the passage of hydrogen into the jar, the palladinized asbestos may be heated and combustion of oxygen may take place at the same time. Thus the last trace of oxygen in the jar is burnt off by the electrical discharge.

After 20 minutes, the indicator tube becomes perfectly colourless ; the suction and the gas supply needle-valves are then closed. Hydrogen cylinder connection and the electric connections are discontinued. The jar is then placed in the incubator when the indicator tube containing alkaline methylene-blue remains colourless, indicating a complete anaerobiasis. After use, the indicator tube is left in place and may be used subsequently.

When hydrogen is passed from Kipp's apparatus, it should be purified by passing through Woulfe's bottles, containing (i) 10 per cent of lead acetate to remove H_2S, (ii) a 10 per cent silver nitrate solution to absorb arseniuretted hydrogen. (iii) a mixture of pyrogallic acid and caustic soda to remove traces of oxygen.

8

Bacterial Communication and Interaction

Bacteria communicate with other members of their community through the secretion and perception of small chemical cues or signals. The recognition of a signal normally leads to the expression of a large suite of genes, which in some bacteria are involved in the regulation of virulence factors, and as a result, these signaling compounds are key regulatory factors in many disease processes. Thus, it is of interest when studying pathogens to understand the mechanisms used to control the expression of virulence genes so that strategies might be devised for the control of those pathogens. Clearly, the ability to interfere with this process of signaling represents a novel approach for the treatment of bacterial infections. There is a broad range of compounds that bacteria can use for signaling purposes, including fatty acids, peptides, *N*-acylated homoserine lactones, and the signals collectively called autoinducer 2 (AI-2). This chapter will focus on the latter two signaling systems as they are present in a range of medically relevant bacteria, and here we describe assays for determining whether an organism produces a particular signal and assays that can be used to identify inhibitors of the signaling cascade. Lastly, the signal detection and inhibition assays will be directly linked to the expression of virulence factors of specific pathogens.

Bacterial Cell-Cell Communication

It is now recognized that bacteria regulate high-density phenotypes, such as biofilm formation, and phenotypes that are important for a number of pathogenic bacteria, through signal-based regulatory systems.

In general, these systems function via extracellular signals that bacteria use to assess their local population density. When these signals reach sufficient concentrations (e.g., at high cell densities), a feedback regulatory loop is induced, resulting in a very rapid expression of phenotypes in the population of cells. The best studied of these regulatory loops are the acylated homoserine lactone (AHL) system in Gram-negative bacteria and the autoinducer 2 (AI-2) system, first identified in the genus *Vibrio* and subsequently found in a broad range of Gram-negative and Gram-positive bacteria.

AHL-mediated density-dependent signaling (or quorum sensing) occurs when a threshold concentration of the AHLs (produced by the product of the *synthase I* gene) accumulates to a sufficient level to interact with receptor proteins (produced from the products of the *R* gene), which then act as transcrip-tional regulators by binding to the promoter of target genes. *Pseudomonas aeruginosa* is a particularly aggressive and problematic pathogen because it produces a number of virulence factors (extracellular proteases, toxins, siderophores), many of which are regulated by AHL-mediated quorum-sensing pathways, and thus *P. aeruginosa* has become the model bacterium for AHL-mediated gene regulation. Several virulence factors in *P. aeruginosa* are AHL-regulated, including elastase, pyocyanin, rhamnolipids, alkaline protease, and exotoxin A. Importantly, the colonization of surfaces and biofilm formation are also regulated by the AHL system in *P. aeruginosa*. For example, AHL-deficient mutants of *P. aeruginosa* form aberrant biofilms and are more easily removed from surfaces.

Moreover, AHL-deficient strains of *P. aeruginosa* are not able to establish chronic lung infections in mouse models of cystic fibrosis infection or ocular disease, which was presumably due to reduced biofilm formation and virulence factor expression. These findings have been replicated with AHL mutants in other pathogenic bacteria including *Burkholderia cepacia* and *Serratia marcescens* (*liquefaciens*). AHL systems in *S. marcescens* control surface colonization, including attachment, surface motility and biofilm development, and expression of other virulence factors [e.g., proteases]. We also have recently demonstrated that the AHL regulatory system in *S. marcescens* controls genes essential for attachment and invasion into corneal epithelial cells.

The AI-2 quorum-sensing system is found in a variety of Gram-negative and Gram-positive bacteria and has been shown to regulate a number of phenotypes, including virulence factor expression and biofilm formation in *Streptococci*, virulence factor expression in enterohemorrhagic *Escherichia coli* (EHEC), and virulence factor expression and biofilm formation in *Vibrio vulnificus* and *Vibrio cholerae*. There are at least 12 pathogenic *Vibrio* species recognized to cause human illness. Among the non-cholera *Vibrio* species are *Vibrio parahaemolyticus*, which is responsible for gastroenteritis and wound infections, and *Vibrio alginolyticus*, which causes wound infections. The most commonly isolated species is *V. vulnificus*, which is of great medical importance due to the high mortality rate associated with infections. Because of the introduction of antibiotics for the treatment of cholera, the occurrence of drug-resistant *V. cholerae* strains is being reported with increasing frequency. Ninety-eight percent of clinical O1 isolates showed resistance to one or more antibiotics, whereas 50% of the environmental isolates showed resistance to one to three drugs, and one clinical and one environmental isolate showed resistant to seven antibiotics.

The trend of acquisition of resistance to antibiotics has increased for many pathogenic bacterial species. Yet, the number of therapeutically useful compounds that emerge is continuously decreasing, and only one new class of antibiotic has been introduced in the last 40 years. Only the development of new classes of antimicrobials with novel mechanisms of action can address the drug resistance problem we face today. Schmidt states "a prioritized goal in the search for new targets is to identify and inactivate disease-causing functions instead of focusing on killing. Since these factors have no vital functions for microbes, there is less pressure to develop resistance, which would ultimately reduce global selective pressure and prevent the acceleration and dissemination of resistance." Thus, quorum-sensing systems such as the AHL and AI-2 regulatory systems represent novel targets for the control of biofilm formation and virulence factor production by bacteria.

Materials

Detection and inhibition of N-acylated homoserine lactone quorum sensing

ABt medium

1. Solution A—in 100 mL of distilled water (dH_2O), 4 g of $(NH_4)_2SO_4$, 6 g of $Na_2HPO_4 \cdot 2H_2O$, 3 g of KH_2PO_4, 3 g of NaCl. Autoclave and store at room temperature.
2. 1 *M* $MgCl_2$.
3. 0.1 *M* $CaCl_2$.

4. 0.01 *M* $FeCl_2$ (make fresh as Fe oxidizes rapidly into insoluble hydroxides).
5. Thiamine, 1 mg/mL.
6. Solution B—to 1 L of sterile dH_2O, add 1 mL of 1 *M* $MgCl_2$, 1 mL of 0.1 *M* $CaCl_2$, 1 mL of 0.01 *M* $FeCl_2$, and 2.5 mL of thiamine at 1 mg/mL. Store at room temperature.
7. 20% glucose, filter sterilized. Store at room temperature.
8. 20% casamino acids, filter sterilized. Store at room temperature.
9. To prepare the ABt medium, add one part of Solution A to nine parts of Solution B. Add glucose and casamino acids to a final concentration of 0.5%. Store at room temperature away from direct sunlight.

Signals and antagonists

1. Stock solution of *N*-3-oxo-hexanoyl homoserine lactone (OHHL), 10 μM dissolved in dimethylsulfoxide (DMSO). Store at –20°C.
2. High-performance liquid chromatography (HPLC) grade dichloromethane (DCM).
3. Stock solution of sample extract, resuspended to 5 mg/mL in 100% ethanol.

Inhibition of the AHL-regulated phenotype of elastase production

1. Elastin-Congo Red: Elastin-Congo Red buffer, 0.1 *M* Tris–HCl (pH 7.2), 1 m*M* $CaCl_2$, Elastin-Congo Red, 20 mg of powder per reaction tube containing 1 mL of Elastin-Congo buffer. The powder is insoluble in the buffer, and hence, each reaction tube has to be measured separately.

Detection and inhibition of AI-2 quorum sensing

1. Autoinducer bioassay (AB) medium: Solution A—0.3 *M* NaCl, 0.05 *M* $MgSO_4$, 0.2% vitamin-free casamino acids. Make 1 L of solution A by adding the reagents above to the indicated final concentrations in dH_2O. Adjust pH to 7.5 with KOH, autoclave, and cool. Add to solution A, 10 mL of 1 *M* potassium phosphate buffer (pH 7.0), 10 mL of 0.1 *M* L-arginine, 20 mL of glycerol, 1 mL of riboflavin (10 μg/mL) and 1 mL of thiamine (1 mg/mL).
2. Hemolysin buffer: 10 m*M* Tris–HCl (pH 7.2), 0.15 *M* NaCl, 10 m*M* $MgCl_2$, 10m*M* $CaCl_2$, 0.2% (w/v) gelatin. Add the reagents above to the indicated final concen-tration in dH_2O. Sterilize by autoclaving.
3. Lauria Broth 20: To 1 L of dH_2O, add 10 g of tryptone, 5 g of yeast extract, and 20 g of NaCl. Autoclave.

Methods

AHL signal, inducer, or antagonist extraction and preparation

In many cases, one may be interested in recovering either quorum-sensing signals from particular bacteria or potential quorum-sensing inhibitors (QSIs) from eukaryotic organisms or from microorganisms. For the isolation of QSIs from eukaryotes, whole materials are normally extracted for active compounds, whereas the isolation of signals or antagonists from bacteria is performed on supernatants of spent cultures. This section describes both procedures.

Extraction of whole materials or supernatants

1. For whole materials, place material to be extracted in fresh DCM and shake.
2. Transfer the material to fresh DCM and repeat for a total of three extractions (skip to step 6).
3. For the extraction of spent supernatants of microorganisms, remove the cells by centrifugation and decanting off the supernatant.
4. Add the DCM in a 2:1 ratio to the supernatant in a separatory funnel, shake well, and let the mixture separate into two phases.
5. Collect the lower solvent fraction and add fresh DCM to the aqueous phase, shake, and reseparate. Repeat for a total of three extractions.
6. Combine the solvent fractions from the three extractions.
7. Evaporate off the solvent using a rotary evaporator.
8. When dry, resuspend the extract in a small (e.g., 1 mL) volume of a suitable organic solvent such as ethyl acetate or DCM and transfer to a clean vial that has been weighed.
9. Evaporate off the solvent, determine the amount of extracted material by reweighing the dried vial, and resuspend to a concentration of 5 mg/mL in ethanol or DMSO.

Minimum inhibitory concentration (MIC) determination for antagonists

Prior to testing of extracts, particularly for extracts of whole organisms, it is useful to determine whether the extracted material affects the growth of the monitor strains. This information can then be used to set the maximum non-inhibitory concentration to be used in the signal induction or inhibition assays.

1. Prepare an overnight culture of the assay strain, *E. coli* MT102 pJBA132 (*luxR*-P_{luxI}-RBSII-*gfp*(ASV)-T_0-T_1) by growing the organism in 2 mL of ABt medium, with 6 μg/mL tetracycline at 37°C with shaking (200 rpm).

2. The following morning, add 100 μL of fresh medium to the wells of a 96-well microtiter plate. To the wells of the top row, add 200 μL total.
3. Add the extracted material to the first row to a maximum of 200 μg/mL.
4. Perform doubling dilutions of the material from the top row to the second to last row, leaving the bottom row without extract as a control row.
5. Dilute the overnight culture 1:100 into fresh medium, with or without antibiotic, respectively, and aliquot 100 μL into the wells of the 96-well microtiter plate containing the extracts for testing.
6. Incubate at 37°C and take regular readings for up to 24 h to determine growth yield and to calculate the MIC values.

AHL induction/inhibition detection

For testing for the presence of inducers or repressors of AHL-based quorum sensing, the same *E. coli* monitor strain can be used. To detect inducers, one can use directly the monitor strain, which does not produce its own AHL signal. Whereas this monitor strain works well for most AHL molecules, it is not sensitive to C4-AHLs and another strain must be used, and the method for testing for the C4-AHLs is described elsewhere. To test for inhibition of signaling, one must use the monitor strain in the presence of both the test substance and exogenously added AHL, which in this case is the 3-oxo-6C-AHL.

1. Prepare an overnight culture of the assay strain, *E. coli* MT102 pJBA132 (*luxR-P*$_{luxI-}$-RBSII-*gfp*(ASV)-T_0-T_1), by growing the organism in 2 mL of ABt medium, with 6 μg/mL tetracycline at 37°C with shaking (200 rpm).
2. (a) To test for induction of the AHL-mediated quorum-sensing monitor, the following morning, add 100 μL of fresh ABt medium to the wells of a 96-well microtiter plate. To the wells of the top row, add 200 μL total. (b) To test for the presence of AHL inhibitors in the extract, the following morning, add 100 μL of fresh ABt medium with 3-oxo-C6-AHL added to a final concen-tration of 20 n*M* to the wells of a 96-well microtiter plate. To the wells of the top row, add 200 μL total volume.
3. Add the extracted material to the first row with the maximum concentration being equal to twice the MIC as determined above or 200 μg/mL, which ever is lower.

4. (a) When testing for the presence of AHL inducers, perform doubling dilutions of the material from the top row to the second to last row, leaving the bottom row without extract as a control row. To the last row, add 3-oxo-C6-AHL added to a final concentration of 20 n*M* as a positive control for induction. (b) When testing for the presence of AHL inhibitors, perform doubling dilutions of the material from the top row to the second to last row, leaving the bottom row without extract as a control row. This last row will serve as a control for maximum induction in the absence of inhibitor.
5. Dilute the overnight culture 1:100 into fresh medium, with or without antibiotic, respectively, and aliquot 100 μL into the wells of the 96-well microtiter plate containing the extracts for testing.
6. Incubate at 37°C and take regular readings for up to 24 h to determine growth yield (optical density at 600 nm—OD_{600}) and quorum-sensing-regulated Gfp expression (excitation 475 nm, emission 515 nm) to calculate quorum-sensing induction and inhibition.
7. Normalize the Gfp expression by dividing the Gfp expression by the OD_{600} to correct for any differences in cell number. The inhibition can also be visualized using fluorescence microscopy.

Inhibition of AHL quorum-sensing phenotypes

To verify that a particular extract or compound is active as an inducer or inhibitor of AHL-based quorum sensing in a relevant pathogen, it is important to validate the bioassay results with a phenotype relevant to the particular pathogen. For this purpose, the expression of elastase, the product of the gene *lasB* of *P. aeruginosa*, will be exemplified. This protease has been selected because elastase expression has been demonstrated to be regulated by quorum sensing in *P. aeruginosa* and the assay is simple to perform. The assay can be performed using either a wild-type strain of *P. aeruginosa* or a quorum sensing mutant strain that does not produce its own signal (e.g., a *lasI*/*rhlI* mutant strain). Presented below is a protocol using a wild-type strain, which is optimum for testing for inhibitors because the elastase production is expressed from native promoters and by endogenously produced signal.

MIC determination

1. Prepare an overnight culture of *P. aeruginosa* by growing the organism in 2 mL of ABt medium at 37°C with shaking (200 rpm).

2. The following morning, add 100 μL of fresh medium to the wells of a 96-well microtiter plate. To the wells of the top row, add 200 μL total volume.
3. Add the extracted material to the first row to a maximum of 200 μg/mL.
4. Perform doubling dilutions of the material from the top row to the second to last row, leaving the bottom row without extract as a control row.
5. Dilute the overnight culture 1:100 into fresh medium and aliquot 100 μL into the wells of the 96-well microtiter plate containing the extracts for testing.
6. Incubate at 37°C and take regular readings for up to 24 h to determine growth yield and to calculate the MIC values.

Induction or inhibition of elastase activity in P. aeruginosa performed with some modifications

1. Grow the *P. aeruginosa* in ABt at 37°C with shaking (200 rpm).
2. The following morning, dilute the culture 1:100 into fresh medium, either add extract at half the MIC dose, as determined above or 200 μg/mL, or add an equal volume of solvent to the control culture and incubate at 37°C with shaking (200 rpm).
3. At hourly intervals, collect 1.5 mL of supernatant and also determine the OD_{600}. Filter out the cells from the 1.5-mL aliquot using a 0.2-μm filter and store at –20°C.
4. Once samples have been collected, thaw on ice. Whilst thawing, prepare tubes with 1 mL of Elastin-Congo red buffer and 20 mg of Elastin-Congo red.
5. Add 100 μL of sterile supernatant to the Elastin-Congo red tubes and incubate at 37°C for 18 h with agitation.
6. Stop the reaction by the addition of EDTA to a final concentration of 12 m*M*. Centrifuge at 16,000 × *g* to remove insoluble Elastin-Congo red.
7. Collect the supernatant and measure the OD_{495}. A blank, with Elastin-Congo red and buffer, but without sample, should be used to subtract out the background. Results should be normalized to correct for differences in cell density.

Detection of AI-2 activity or inhibition

AI-2 activity is commonly detected by the use of AI-2 bioassay developed over a decade ago. *Vibrio harveyi* BB 170 (sensor 1^-, sensor 2^+) has a null mutation in the gene for sensor 1 and thus is a

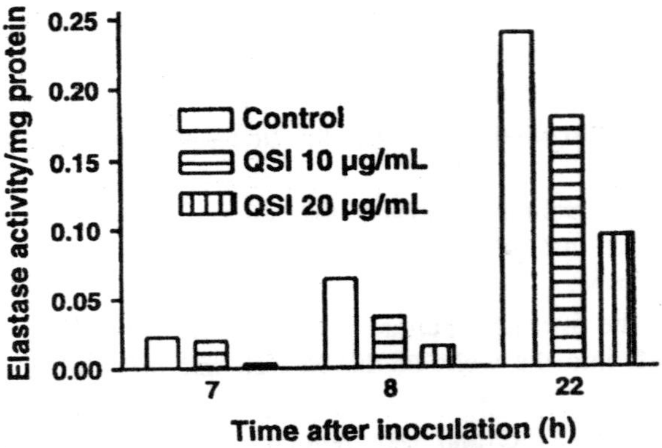

Fig. 8.1. Elastase activity of Pseudomonas aeruginosa in the presence and absence of a quorum-sensing inhibitor (QSI).

bioluminescent reporter strainforAI-2 activity. *V. harveyi* BB152 (autoinducer 1^-, autoinducer 2^+) has a null mutation in the gene for autoinducer 1 synthase (AI- 1) and produces only AI-2, and thus serves as a positive control for assays of AI-2 production. For the generation of cell-free supernatants, *V. harveyi* strains were grown in AB medium overnight at 30°C with shaking at 200 rpm and cell-free supernatants prepared as below.

Collection of AI-2 signal

1. For generation of cell-free supernatants, test cultures are inoculated at a 1:100 dilution and the cultures grown to late stationary phase. Growth is followed spectrophotometrically and aliquots removed at various time points for the preparation of cell-free supernatants.
2. Cell-free supernatants are prepared by centrifugation of culture aliquots for 5 min at 6000 × *g* followed by filtration of the supernatant through a 0.22-μm filter (Pall acrodisc supor syringe filter) and collection of the supernatant in a sterile 1.5-mL Eppendorf tube.
3. Cell-free supernatants can be frozen and stored at –20°C until used.

Detection of AI-2 activity/inhibition

To test for the presence of AI-2 activity, grow the *V. harveyi* reporter strain BB170 for 16 h at 30°C with shaking in AB medium.

1. Dilute cells 1:5000 into 30°C pre-warmed AB medium.
2. Prepare microtiter plates by adding 90 μL of the diluted suspension to wells of 96-well microtiter plates and then add 10 μL of either medium (for negative controls) or supernatants to be tested.

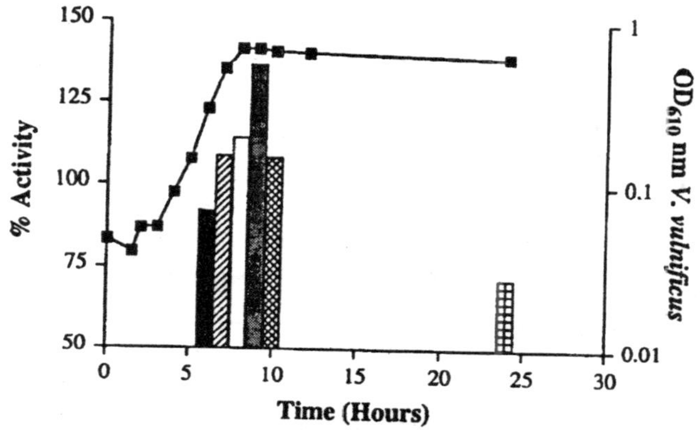

Fig. 8.2. Production of autoinducer 2 (AI-2) signal by Vibrio vulnificus during different phases of growth.

3. Incubate the microtiter plates at 30°C with shaking at 175 rpm. Hourly determi-nations of the total luminescence as well as cell density are quantified.
4. Report activity as the percentage of activity obtained from *V. harveyi* BB152 cell-free supernatant.
5. To test for inhibition of AI-2-mediated cell–cell signaling, add the diluted monitor strain to the plates as indicated in steps 1–3, where the supernatant added is from the positive control strain, *V. harveyi* BB152. Add the antagonist from the stock solution, as extracted above. Again, it is advisable to first perform MIC determinations of the inhibitor extract as described above to avoid effects on growth. Representative data show the timing of AI-2 production during the growth of *V. vulnificus*.

Inhibition of hemolysin assay with signal antagonist

To test for inhibition of hemolysin activity by signal antagonists, routinely grow the test cultures overnight. The next morning, 10 mL of fresh medium is inoculated with a 1:100 dilution of the overnight culture in a sidearm flask and incubated at 37°C with shaking (200 rpm) and growth measured at OD_{610}.

Antagonists are added to the cultures during log phase (OD_{610} is approximately 0.3). DMSO is added to control samples to account for the solvent volume added, and growth should be followed spectrophotometrically to ensure the concentrations of antagonist compounds used are not growth inhibitory. Cell-free supernatants should be collected across the growth curve.

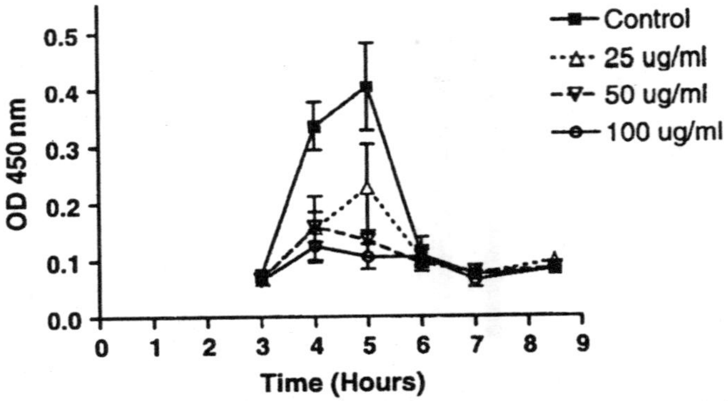

Fig. 8.3. Inhibition of the autoinducer 2 (AI-2)-mediated production of hemolysin in Vibrio vulnificus.

1. Wash 4 mL of sheep blood three times with five volumes (20 mL) of PBS (pH 7.2). Centrifuge at 1140 × *g* for 5 min and resuspend in five volumes (20 mL) of PBS.
2. Incubate aliquots of cell-free supernatants (200 μL), 600 μL of hemolysin buffer, and 200 μL washed red blood cells at 37°C for 30 min.
3. Remove intact red blood cells by centrifugation and determine the OD_{450} of the supernatant.

Notes

1. ABt medium can be stored at room temperature, but should not be more than 1 month old when using.
2. It is important to use high-quality solvents or to distill them because many commercial grade solvents contain plasticizers that are concentrated during the extraction procedure. These concentrated compounds can be toxic to the indicator strains or can interfere with the interpretation of results.
3. It is preferable to use a solvent that is miscible in aqueous media, such as ethanol or DMSO for the bioassay testing. However, if the extract does not dissolve well in ethanol, it is possible to evaporate off the ethanol and resuspend the sample in an organic solvent and then dilute into ethanol or DMSO prior to adding to the bioassay medium. DMSO is not a volatile solvent and hence if the sample does not dissolve in DMSO, it is not easy to remove the DMSO to try another solvent.
4. Some materials, such as plants, can lyse in organic solvents. This is only an issue if the experimental design intends to separate

surface active or presented compounds from intracellular compounds. If the former is the case, then different organic solvents and extraction times should be tested to identify conditions that do not lyse the tissues.

5. It is best to grow the organisms in defined minimal medium because some components of complex media interfere with the AHL assays and can give false-positive results. Typically, 100 mL of culture supernatant is sufficient for detection of signals or antagonists.
6. At this point, the maximum concentration has been halved due to the addition of the cells.
7. This method will not identify the chemical species that induces or inhibits the quorum-sensing monitor strain. There are other methods that can aid in this, such as thin layer chromatography or gas chromatography–mass spectroscopy methods.
8. The timing of addition relative to sampling is partly dependent on the stability of the signal or antagonist in the presence of bacteria. Therefore, if no effect is observed with a single dose, it may be necessary to repeat the dose during growth or to time the addition of the extract more closely with the onset of elastase production.
9. Filtration typically results in the loss of some sample volume, so a slight excess is collected.
10. Normalization for cell density can be done by dividing the OD_{495} by either the OD_{600} or by protein concentration.
11. It is important to note that 5 g/L of glucose has been shown to inhibit biolu-minescence in *V. harveyi*; thus care must be taken to ensure the growth medium used for the test strains does not contain high levels glucose. The cell-free supernatant can be made glucose depleted by growing *V. harveyi* strain MM30 (AI-2^-) in the supernatant prior to testing. By contrast, low levels of glucose may actually increase bioluminescence and lead to false positives.
12. It is important that the cell-free supernatants be neutral in pH as acidity can interfere with the assay.

Bacteria–Host Cell Interactions

Electron microscopy of bacterial pathogens and interactions between bacteria and host cells and tissues provides valuable insights into structural and molecular properties and processes involved in pathogenesis. Applications for electron microscopy in bacterial pathogenesis range from discovering etiologic agents and following chronological events during infections by conventional examination of clinical samples to assessing molecular host–cell responses to infection and in situ interactions between receptors and ligands using specific immune-labeling techniques. This chapter focuses on techniques for preparing samples of bacteria and host cells for conventional transmission (TEM) and scanning electron microscopy (SEM) and use of luminescent nanocrystals or "*quantum dots*" as specific probes for correlative light and electron microscopy. Conventional TEM and SEM are well established tools for high resolution examination of structural effects and chronological events associated with bacterial infections. The recent development of quantum dots as physiological and immunological probes in biology has provided a powerful technique for bridging fluorescent analyses of fixed and live material with preparation and examination by TEM and SEM.

Historical Overview

Ultrastructural analyses by TEM and SEM are versatile and integral tools for examining and characterizing infectious cycles and host responses to bacterial pathogens at the tissue, cellular, sub-cellular, and molecular levels. Utilizing focused electron beams with wavelengths less than 100 pm, electron probe microimaging augments biochemical analyses by providing structural and spatial information at resolution 2–3 orders of magnitude greater than that achievable by standard far-field optical microscopy. Together with optical imaging of fixed and live preparations, TEM and SEM provide numerous techniques for investigating and analyzing chronological and structural relationships of interactions between bacteria and their hosts and environments

Protocols for conducting such investigations vary considerably according to particular aspects of experimental design and desired outcome. Examples include imaging techniques, such as negative staining for structural or immunological analyses of bacteria, rapid freezing and freeze- substitution for preservation and immobilization of fine structures and aqueous solutes, microwave processing for rapid diffusion and infiltration of fixatives and embedding resins, variable pressure SEM and containment chambers for examining hydrated samples, and enzymatic or immunological labeling for locating and identifying specific target molecules. Such techniques are commonly used to assess bacterial morphology and phenotypic expression of surface and extracellular components such as flagella, pili, fimbrae, and capsules. Furthermore, TEM and SEM can reveal the chronology of microbial cell cycles, phase changes, sporulation, and cytopathic effects on host cells and tissues, such as cytoplasmic and cytoskeletal rearrangements, plasma membrane ruffling and pedestal formation, organelle trafficking and vesicle fusion, and cell lysis, necrosis, apoptosis, and autophagy. Despite the diversity of TEM and SEM applications and approaches to microbial investigations, most protocols share common processes to address properties and requirements inherent with electron probe analyses of biological material, such as a high vacuum environment and both the damaging effects and exploitable attributes of focused cathode-ray irradiation. This chapter provides descriptions of conventional preparative and immunological-labeling techniques with which many questions can be addressed and modifications adapted. For in-depth review of theoretical and practical applications of electron microscopy, readers are directed to any of several excellent volumes.

Because of the broad nature of microbial samples prepared for EM, this chapter cannot describe methods to address all types of ultrastructural studies. Rather, this chapter provides methods for preparing samples that are either cultured on or subsequently attached to a rigid or semi-rigid substrate. Examples of such substrates are silicon chips, Thermanox cover slips, Aclar film, filter membranes, or agar-based culture medium. Adherence to a firm substrate is advantageous for positioning TEM samples for sectioning along planes of interest, for maintaining the integrity of biofilms and host cell layers, and for interpreting structures and events that may be polarized with respect to basal, lateral, or apical cell surfaces. In many instances, such layers are appropriate and useful in vitro models for the native associations of bacteria and host cells. Microbial and cell adherence to such substrates can be promoted, if necessary, by pre-coating the substrate with a biologically relevant substance, such as fibronectin, vitronectin, collagen, or poly-L-lysine. It should be noted, however, that many situations exist in which adherence to a surface is not advantageous and use of centrifugal pellets, for example, may be preferable.

During all steps of sample preparation, care should be taken to assure that the samples, reagents, containers, and tools are kept clean and free of particulates and volatile contaminants. Dust, fingerprints, trace salts, and other materials in solution degrade image quality and can damage or destroy samples and instrumentation. To minimize problems with contamination, several preventative steps are advisable, including (1) acquiring high quality water and reagents, (2) using pre-cleaned or sterile plasticware and thoroughly cleaning and rinsing all glassware, (3) filtering buffers, stains, and sample rinse water with 20–200 nm filters prior to use, (4) handling samples, tools, and sample holders with protective gloves, and (5) reducing dust by filtering incoming ventilation and controlling foot traffic into preparative and instrument areas. Furthermore, several reagents and resins used in sample preparation are highly toxic, carcinogenic, and/or radioactive. Therefore, material safety data and recommendations from the suppliers should be obtained, understood, and followed for safe handling and disposal of all potentially hazardous reagents.

Conventional TEM

This section provides preparative methods for examining thin sections of microbial material. These techniques are suitable for samples that may or may not include specific staining or labeling steps for identifying or highlighting structures or molecules of interest. It is presumed that certain equipment including a fume hood, ultramicrotome and diamond or glass knife, and transmission electron microscope are available for use.

Preparative steps involve fixing with cross-linking agents, contrasting with electron-dense compounds, dehydrating, embedding, and sectioning. Typically, primary fixation involves immersion in buffered glutaraldehyde or a mixture of glutaraldehyde and paraformaldehyde. These reagents react with and can cross-link free amino groups. If needed, the primary fixative can also be used for perfusion of tissues. As the rate of diffusion and thus fixation is dependent on the biochemical nature of the material and physical properties such as temperature and sample thickness, the length of time required is variable and largely unpredictable. Organisms with thickened, resistant, and/or hydrophobic cell walls such as Gram-positive bacteria, spores, or mycobacteria require extended fixation or enhanced processing, such as microwave irradiation or both. Experimental assessment of viability during fixation can be useful for determining the necessary length of treatment. Membranes and storage lipids are stabilized by cross-linking with oxidative osmium species. Osmium treatment both reduces the quantity of lipids extracted during preparation and provides contrast for lipophillic structures. Contrast is often enhanced by subsequent staining with metallic salt solutions, such as uranyl acetate and lead citrate. To enable thin sectioning, the samples are dehydrated in organic solvents and infiltrated with epoxy or acrylic resins, which are then polymerized to provide support and minimize compression artifacts during sectioning. As with fixation steps, the duration period and processing techniques for dehydration and embedment may require adjustment depending on types of samples and viscosity of the resin.

Conventional SEM

For conventional SEM, samples must also be cross-linked and dehydrated during preparation. Treatments with aldehydes and osmium oxides and solvent dehydration can often be performed identically for both TEM and SEM. Once fully dehydrated in 100% organic solvent, the SEM samples are dried and mounted. To minimize shrinkage during drying, intermediate drying agents are used. This can be an organic compound, such as hexamethyldisilazane, or more typically the agent is liquid carbon dioxide exchanged within a critical point dryer. Once dried, the samples are coated with a conductive metal or carbon before examination or examined without coating. Coating, usually performed in a magnetron or ion beam sputterer, can provide enhanced conductivity, some degree of protection against beam damage, and perhaps most importantly, increased production of secondary electrons for imaging. Although SEM offers several modes of imaging for biological samples, typically samples are examined using secondary electrons for cell and microbial surface topography and backscattered electrons for compositional contrast of immune probes and labels.

Immune Labeling with Quantum Dots

Immune labeling of biological material for electron microscopy is a well- established and powerful tool for localizing and visualizing macromolecules and molecular complexes with nanometer scale resolution. In recent years, considerable efforts have been undertaken to develop and utilize techniques for correlative microscopy involving assessment of equivalent samples by fluorescent (including single and multi-photon confocal microscopy) and electron microscopy. Because of the inherent differences between light and electron beam imaging, relatively few techniques have been developed that can utilize the same specific labeling reagents. These include (1) use of fluorescein-conjugated probes for fluorescence microscopy, which can be subsequently labeled with anti-fluorescein antibodies and electron-dense metals, (2) use of the green fluorescent protein family of expressed markers that can also be labeled with antibodies and electron-dense metals, the ReASH system, which utilizes a genetically engineered tetracysteine motif and a bi-arsenical fluorescent dye that can photoactively deposit diaminobenzidine for TEM detection of the expressed protein tag, and (4) the use of fluorescent CdSe nanocrystals known as "*quantum dots*," which are photoluminescent, phosphorescent, and electron dense. Among these options, the quantum dots are unique in that they enable fluorescent examination of labeled live or fixed samples and subsequent preparation of the same samples for electron microscopy without further manipulation of the label. This was recently described for simultaneous detection of several quantum dot probes by fluorescence and TEM of sections by the wavelength of emission and size class of particles, respectively. To date, there are no reports of SEM studies using quantum dots as immune labels. However, phosphorescent emissions similar in wavelength to photoluminescent emissions have been detected by SEM with a cathodoluminescence spectrometer. Studies are underway to determine whether cathodoluminescent properties of quantum dots will be adaptable for immune labeling of biological material for SEM. In the meantime, the CdSe core and ZnS coating of quantum dots provide for compositional contrast using backscattered electron detection. Labeling and imaging methods described below enable direct detection of the quantum dots by TEM and low-voltage SEM.

Materials

Conventional TEM

1. Hank's-buffered salt solution (HBSS).
2. 25% aqueous glutaraldehyde, EM grade, in nitrogen-filled ampules.
3. 16% aqueous paraformaldehyde, EM grade, in nitrogen-filled ampules.
4. Sodium cacodylate (dimethyl arsenate).
5. 4% aqueous osmium tetroxide, EM grade.
6. Potassium ferrocyanide.
7. 70 and 100% (dehydrated) ethanol.
8. Spurr's embedding medium.
9. Uranyl acetate.
10. Lead citrate.
11. 10 *N* sodium hydroxide.
12. Disposable serological pipets.
13. Disposable transfer pipets.
14. Powder-free nitrile or other appropriate gloves.
15. Thermanox cover slips, 13 mm.
16. Embedding capsules or molds.
17. Copper EM grids, 200–300 mesh.
18. Grid box(es).

19. Laboratory film such as Parafilm.
20. Syringe filters, 20 nm Anotop.
21. Disposable Leur-Lock syringes, 10 mL.
22. Fine tweezers, such as anti-capillary tweezers.

Conventional SEM

1. Same as items 1–7 and 12–14 in TEM.
2. Silicon chip specimen mounts.
3. 24-Well tissue culture plates.
4. SEM sample stubs compatible with the microscope.
5. Chromium, iridium, or other appropriate sputtering target.
6. Carbon sputtering target.

Immune labeling with quantum dots

1. Phosphate-buffered saline (PBS).
2. Immunoglobulin-free bovine serum albumin (BSA).
3. Primary antibodies or anti-serum, for example, rabbit anti-serum.
4. Secondary antibody conjugate, for example, goat anti-rabbit 655 nm quantum dots.
5. Parafilm or similar laboratory film.
6. Anti-capillary tweezers (Ted Pella, Inc).

Methods

Conventional TEM

This section describes steps for fixating and contrasting, dehydrating, embedding, sectioning, and mounting samples for TEM. Many variations on these steps have been described depending on types of samples used, experimental goals, and personal preference. This set of methods was used successfully to demonstrate stages of infection of primary human neutrophils with clinical isolates of group A *Streptococcus*, and methicillin-resistant *Staphylococcus aureus* (MRSA), and lymphocytes by *Borrelia burgdorferi*. The samples of neutrophils were allowed to adhere to 13-mm Thermanox cover slips in the wells of a 24-well tissue-culture plate prior to exposure to bacteria. In contrast, 10^5 B cells and 10^6 spirochetes were co-incubated within liquid growth medium suspensions and then processed for TEM as pellets. In these examples, the time periods for each step are relatively long and reflect allowance for slow diffusion of reagents through Gram-positive cell walls, and pellets of cells and spirochetes, respectively. In all processing steps, care must be taken to avoid unintentional drying of samples. When necessary, centrifugation and resuspension should be conducted as gently as possible to minimize mechanical damage. Furthermore, TEM grids must be kept clean and handled very carefully, grasping only at the edges with very fine tweezers.

Fixation

1. Immerse samples in at least 10 volumes of Karnovsky's fixative containing 2.5% glutaraldehyde and 4% paraformaldehyde in 0.1 *M* sodium cacodylate buffer, pH 7.2 for 2 h.
2. Wash the samples twice for 30 min each time by replacing the fixative with cacodylate buffer.
3. Post-fix the samples by replacing the wash buffer with a mixture containing 1% osmium tetroxide and 0.8% potassium ferrocyanide in 0.1 *M* cacodylate for 2 h.
4. Wash the samples once for 30 min in cacodylate and twice in deionized or distilled water.

In-block staining

1. Stain the samples by immersion in 1% uranyl acetate in water for 2 h at room temperature.
2. Wash the samples twice for 30 min each in water.

Dehydration and embedment

1. Dehydrate the samples by immersion for 1 h each in 70, 100, and 100% ethanol.
2. Prepare embedding resin according to the manufacturer's instructions and mix thoroughly.
3. Infiltrate samples with embedding resin by immersing in 1/3 resin and 2/3 ethanol for 4 h, 2/3 resin and 1/3 ethanol for 4 h, and 100% resin for 8–16 h twice.
4. After two cycles in 100% resin, transfer the samples to resin-filled embedding capsules or other suitable molds designed for curing at elevated temperature. Orient the samples for the optimal position and angle of sectioning. For centrifuged material, this is typically not an issue due to the random orientation of cells and bacteria within the pellets. For adherent cells on Thermanox cover slips, Aclar, or silicon chips, orient the substrates such that the cells face toward the resin at the preferred angle for sectioning, usually either parallel or perpendicular to the cell layer. It is helpful to place a small paper label within the resin block with identifying information written in pencil or printed in a small font (e.g., 5 point font) for later reference.

5. Carefully place the samples into an oven for curing at 65°C or as recommended by the resin supplier. Curing is temperature dependent. At 65°C, blocks are usually hardened within 24–48 h.

Block face preparation and sectioning

1. Remove the cover slip or any other substrate from the block. The cells should remain in the block, and exposed on the surface.
2. Secure the block in a microtome chuck or small vice. Using single-edge razor blades, carefully trim the block down to a suitable block face for sectioning. This is typically a trapezoid ranging in size from about 0.2–1.0 mm on each side.
3. Using an ultramicrotome and glass or diamond knife, cut sections of suitable thickness.
4. Pick up the sections on clean EM grids. The grids can either be coated with thin films such as carbon, Parlodion or Formvar, or left uncoated. Wick off excess water by gently touching absorbant filter paper to the edge of the grid. Place the grid into a grid storage box and record any necessary identifying information.

Post-section staining

Specialized devices are available for post-section staining that facilitate handling of multiple samples and help provide consistency and reproducibility, for example, Ted Pella, Inc., item number 22510. Traditionally, grids have been stained with droplets as follows.

1. On a freshly uncovered surface of ParaFilm or other laboratory film, pipette 30-μL droplets of Anotop-filtered 1% uranyl acetate in water.
2. Carefully place one grid on each droplet, with the sections facing the droplet. The grids will normally float on the droplet unless a surfactant is present. Cover the droplets and samples to protect against dust and drying. Plastic lids from pipette tip racks work well for this purpose. Allow 10 min for staining. In low-humidity environments, it is prudent to supply a source of hydration, such as a wet gauze sponge or laboratory wipe within the covered space to reduce evaporation.
3. Transfer the grids to 100-μL droplets of filtered deionized or distilled water. Thoroughly wash each grid with water by passing 5–10 mL of filtered water over both grid surfaces dropwise. Carefully wick excess water from each grid using filter paper as above. Allow to dry.

4. Prepare fresh 1% lead citrate stain by adding 0.1 g of lead citrate to a mixture of 9.5 mL of water and 0.5 mL of 10 *N* sodium hydroxide. Vortex the mixture until dissolved. Pass through Anotop filters before use. Alternative preparative methods are also available.
5. Repeat steps 1–3 above using 1% filtered lead citrate instead of uranyl acetate. The grids are then ready to examine.

Conventional SEM

The methods described in this section are applicable to a wide variety of samples.

As with the TEM methods, the time periods for each step are sufficient for most samples of individual cells or cell layers with associated Gram-negative or non-spore-forming Gram-positive bacteria. For bulk samples such as biopsied tissues or bacterial colonies on solid medium, incubation periods should be adjusted to ensure thorough diffusion of reagents into the samples.

Samples for SEM examination are often cells, cell layers, or bulk tissues mounted on rigid or semi-rigid substrates. Several preparative steps are similar to those listed above for TEM. The methods diverge to prepare the samples for viewing exposed surfaces rather than internal structures.

Fixation

1. Apply suspended cells to substrate if necessary. Place the desired number of silicon chips into wells of a 24-well tissue-culture plate, shiny side up. Pipette 30 μL of cell and/or bacterial suspensions that have been washed and resuspended at 10^6 or 10^7/mL of HBSS, respectively, onto the shiny surface of silicon chips. Cover and place in a hydrated CO_2-incubator. Allow to settle for 10–15 min.
2. Gently add 1 mL of fixative containing 2.5% glutaraldehyde, 4% paraformaldehyde, and 0.1 *M* sodium cacodylate buffer, pH 7.2, to each well. Allow 30–60 min for primary fixation. Mammalian cells, cell monolayers, and Gram-negative bacteria fix relatively rapidly. Bulk tissue samples, Gram-positive bacteria (particularly spore-forming bacteria), and hydrophobic mycobacteria may require extended fixation time periods.
3. Wash the samples twice for 15–30 min with 1–2 mL of cacodylate buffer.
4. Post-fix the samples with 1 mL of 1% OsO_4 in cacodylate for 30–60 min.

5. Wash the samples once with cacodylate buffer and twice with water.

Dehydration, coating, and mounting

1. Dehydrate the samples by replacing the water with hour-long series of 70, 100, and 100% ethanol.
2. Dry the samples using a critical point dryer and carbon dioxide.
3. Mount the samples, oriented with the material of interest facing upward, onto sample stubs manufactured for the available scanning EM, using conductive material such as silver, gold, or graphite paint, or double-adhesive carbon disks.
4. Sputter-coat the samples with suitable material such as chromium, iridium, gold-palladium, or carbon. The samples are ready for examination.

Immune labeling with quantum dots

The methods described in this section pertain to labeling antigens located on the surface of cells or thin sections for examination by either transmission or scanning electron microscopy (SEM). Despite being similar to colloidal gold for detection by electron microscopy, quantum dots offer the ability to observe labeling and associated biological processes by fluorescence microscopy prior to fixation for electron microscopy. Although use of quantum dots for biological studies is relatively new, a significant number of primary and secondary conjugates of quantum dots with varying sizes and emission wavelengths are commercially available. Furthermore, quantum dots with reactive linkers are also available for producing direct conjugates of interest, making the dots a convenient, adaptable, and flexible choice as probes for molecules and molecular interactions. As above, the protocol described applies to bacteria and cells adsorbed to a substrate such as silicon chips or Thermanox cover slips. If used with pellets or suspended material, the methods can be modified accordingly. For some applications, it may be advantageous to pre-fix the samples to prevent progression of biological processes during labeling. When labeling live material, active processes such as internalization, intracellular trafficking, and cell division can be controlled as needed with temperature or suitable chemicals such as azide or other metabolic inhibitors. Using quantum dots, common steps such as titrating antibodies, conjugates, and blocking reagents to optimize labeling and detection of targets by fluorescence microscopy generally provide optimal electron imaging as well. Thus, fluorescence microscopy is useful for rapidly assessing the specificity and sensitivity of the labeling experiment before the samples are processed for electron imaging. However, this is not necessarily the case when detergents or solvents are used to permeabilize the cells or bacteria, as cellular ultrastructure can be affected adversely and dramatically by such agents, without noticeable detriment in a light microscope.

Pre-fixation (optional)

1. Wash samples twice with an appropriate physiological buffer such as PBS or HBSS to remove residual medium components.
2. Immerse the samples into PBS containing 4% paraformaldehyde and 0.1% glutaraldehyde. Incubate at room temperature for 15–30 min.
3. Wash sample twice for 15 min with buffer.

Immune labeling

1. Block the samples by immersing in PBS containing 2% (w/v) globulin-free BSA. Incubate at room temperature for 30 min.
2. Replace the blocking solution with primary antibody such as immune or pre-immune mouse serum diluted to an appropriate concentration in blocking buffer. Incubate the mixture for 30 min at room temperature or overnight at 4°C. Determining the optimal concentration and length of incubation may require empirical experimentation.
3. Wash the samples at least twice for 15 min in blocking buffer.
4. Label the samples by immersion into 5 n*M* secondary quantum dot conjugate, such as goat anti-mouse 655 nm quantum dots, in blocking buffer. Incubate for 30 min at room temperature.
5. Wash the samples at least twice for 15 min in blocking buffer. Wash the samples two more times in PBS to remove residual protein from the blocking buffer. The samples are now ready to examine by fluorescence microscopy and to proceed with processing for TEM or SEM.

Imaging by TEM

Quantum dots are readily detectable by TEM in thin section and negatively stained preparations at voltages commonly used for biological material between 60 and 120 kV. No special requirements are needed for imaging.

Imaging by SEM

Quantum dots are detectable by SEM by topographical contrast with secondary electron imaging (SEI), compositional contrast with backscattered electron imaging (BEI), and scanning transmission electron imaging. By SEI, the quantum dots appear as particles on exposed surfaces that can resemble normal biological structures. However, the CdSe core and ZnS coating impart clear compositional contrast against carbon-rich biological material when imaged by BEI or scanning transmitted electron detectors. Such contrast is maximized using relatively low accelerating voltages and carbon sputter coating.

1. Follow steps 1–5 as described above.
2. Lightly sputter coat samples with carbon. The coating thickness should be determined empirically as the minimum thickness that enables viewing without static charge artifact in the microscope. A thickness setting of 20 Å was used for the image.
3. Choose an initial accelerating voltage of 1–3 kV. Check and adjust beam alignment as necessary. Examine the sample in backscatter mode at ×25,000 or greater magnification. Quantum dots should be visible as relatively bright particles on exposed surfaces.

Notes

1. Sample processing for EM involves multiple fluid exchanges. Such exchanges must be sufficiently rapid to prevent samples from drying. For example, fluids can be removed from and immediately replaced for each sample in an experiment rather than removing fluids from all samples before replacement.
2. It is the practice in this laboratory to centrifuge suspended cells at 600–800 × *g* for 1–5 min as necessary. After removing the supernatant, pellets are gently disaggregated into the residual liquid by tapping the tube prior to adding the next volume of liquid. Vortexing and rapid passage through pipette tips are not recommended.
3. Sodium phosphate provides excellent buffering for fixatives at 0.1 *M* and is often used in place of cacodylate. Each buffer system has certain advantages and limitations. Cacodylate is an organic arsenic compound, dimethylarsenate, which is poisonous. Whereas it may provide additional TEM contrast in biological samples, it is a relatively weak buffer at pH 7.2. Alternatively, phosphate requires fewer safety precautions. However, our laboratory has observed precipitants in some biological samples, such as avian erythrocytes, that were attributed to the phosphate buffer.
4. Epoxy resins emit noxious and potentially carcinogenic fumes. Therefore, curing ovens should be located within a fume hood or

designed to be vented away from occupied spaces as with vacuum ovens.

5. Removal of Thermanox cover slips can be facilitated by immersing the cover slip into liquid nitrogen for about 5 s. Prolonged exposure of the resin to liquid nitrogen should be avoided, as cracks can develop within the block. Similarly, silicon chips are removed cleanly after pressing the chip onto dry ice for 10–15 s.
6. The quality of sections are affected by many variables including personal ability, type of microtome, type of knife, flaws in the knife, cutting speed, sample and block integrity, ambient temperature, air circulation, humidity, and others. Section thickness settings on ultramicrotomes are approximate values. Typically, sections are cut between 50–70 nm in thickness. At this thickness, the sections appear silver to light-gold in color and provide good characteristics of stability, contrast, and resolution under typical biological imaging conditions of 80–100 kV. Furthermore, compression artifact during sectioning can be minimized by orienting the block face such that parallel edges of the block face are aligned parallel with the knife blade, and that the wider edge advances across the knife first.
7. In some cases, the SEM substrate can be a TEM grid. More typically, ideal substrates are flat or relatively featureless or both. Pre-cleaned glass or plastic cover slips are commonly used for cell layers and cells that can adhere. Use of pre-cut silicon chips as a substrate is gaining popularity, as these are very flat and featureless and sufficiently conductive to minimize charge buildup in samples. In many cases, cells and bacteria adhere to cover slips and chips without further treatment. If necessary, the substrates can be coated with polypeptides or other agents such as collagen, fibronectin, serum, or polylysine to promote adherence.
8. In our laboratory, backscatter detection of quantum dots has been achieved at accelerating voltages ranging from 0.5 to 10 kV. However, during testing, backscatter contrast decreased steadily as the accelerating voltage was increased beyond 2 kV, presumably due to reduced scattering by the nanocrystal or increased scattering by the substrate or both. If available on the SEM, mixing signals from the secondary and backscattered electron detectors, or adjusting for detection of optimum electron energy levels may enhance image quality with respect to compositional contrast and sample topography.

9

Analysis of Bacterial Population

Comparative genomics is a rapidly expanding discipline revolutionized by the availability of increasing numbers of bacterial genome sequences. DNA microarrays that represent single or multiple bacterial genomes are a cost-effective way of comparing the genome content of large numbers of strains of the same species. *Comparative genomic hybridizations* (CGHs) with microarrays allow determination of the presence or absence of genes based on the level of sequence homology required for hybridization to occur.

CGH analyses can result in broad insights into bacterial genomics, population genetics, and evolution. Moreover, whole-genome microarrays can be used to address specific issues of bacterial pathogenesis including the basis for host specificity, epidemic waves, and the emergence and re-emergence of pathogenic clones. Importantly, the findings of such studies could be used in the rational design of disease-controlling strategies.

Studies on the evolution and pathogenesis of an increasing number of bacterial pathogens including *Escherichia coli* O157:H7, *Mycobacterium* spp., *Helicobacter pylori*, *Yersinia* spp., and *Staphylococcus aureus* have been performed with microarrays, resulting in a greatly enhanced understanding of their biology.

Several different approaches to microarray construction have been developed, including oligonucleotide- or DNA-based arrays on glass slides or nylon membranes. Here, we describe a robust CGH protocol based on the use of a PCR-generated microarray. Detailed descriptions of genomic DNA template preparation, labeling, microarray hybridization, and data analysis are provided.

MATERIALS

Genomic DNA Template Preparation (S. aureus)

1. Edge Biosystems DNA extraction kit. All components are kept at 4°C; spheroplast buffer is stored at –20°C.
2. 100% isopropanol (molecular grade). Chill before use.
3. 70% ethanol in sterile deionized water. Chill before use.
4. Lysostaphin (AMBI products) lyophilized powder is resuspended in nuclease-free water (Stock; 5 mg/mL). Store at –20°C.
5. Agilent 2100 Bioanalyser or equivalent.
6. *Hin*P1I restriction endonuclease. Store at –20°C.

Labeling of DNA

1. 400 ng of genomic DNA templates.
2. Prime-a-gene labeling kit:
   (a) Bovine serum albumin (BSA, 10 mg/mL); nuclease free.
   (b) Unlabeled dNTPs (1.5 m*M*). Store at –20°C.
   (c) DNA polymerase I large fragment (Klenow). Store at –20°C.
   (d) Labeling buffer (5×) including 26.0 A_{260} U/mL random hexadeoxyribonu-cleotides.
3. TE Buffer: 10 m*M* Tris–HCl, 1 m*M* EDTA, pH 7. Store at room temperature.
4. 0.2 *M* EDTA (pH 8.0). Store at room temperature.
5. Cyanine 3-dCTP (Cy3); Cyanine 5-dCTP (Cy5) (NEN-Dupont). Store at –20°C; light sensitive.
6. Sephadex G-50 spin columns.
7. Centri-spin[20] label clean up kit. Store at room temperature.
8. Agilent 2100 Bioanalyser or equivalent to determine genomic DNA labeling efficiency.

Slide Preparation and Pre-Hybridization

1. Microarray hybridization chamber.
2. Hybridization buffer, ULTRAhyb, pre-heated to 68°C.

Hybridization

1. 20× sodium chloride sodium citrate (SSC): 3.0 *M* NaCl, 0.3 *M* Na-citrate, pH 7.0.
2. 2× SSC, 0. 1× SSC, and 0.05× SSC in nuclease-free water. Store at room temperature.

3. 0.1% (w/v) SDS. Store at room temperature.
4. 95% ethanol, molecular grade.
5. Lifterslip.

Scanning of Slides and Data Analysis

1. GenePix 4100A microarray scanner or equivalent and GenePix Pro 6.0 software or similar.
2. GeneSpring software.
3. Microarray validation system.

Phylogenetic Analysis

1. Cluster and TreeView programs.

METHODS

The major steps in the CGH protocol include genomic DNA isolation and labeling, microarray hybridization, scanning of slides and calculation of fluorescence intensity ratios, and data analysis by use of appropriate statistical methods.

Genomic DNA Isolation and Digestion

1. Perform genomic DNA isolation from *S. aureus* test and reference strains using a modification of the Edge Biosystems genomic DNA isolation kit. The modification consists of addition of lysostaphin (100 μg/mL final concentration) to bacterial cells and incubation at 37°C for 10 min prior to DNA extraction.
2. Determine the quality of DNA on a Bioanalyser or equivalent; the $A_{260/280}$ nm absorbance ratio should be >1.8. The procedure outlined is optimal for 400 ng of staphylococcal genomic DNA.
3. Digest genomic DNA with the restriction enzyme *Hin*P1I at 37°C overnight and purify before using as a template for labeling.

Labeling of DNA

1. Thaw all components on ice. Keep the Klenow fragment at -20°C until required.
2. Denature purified, digested genomic DNA in TE buffer at 95°C for 2 min followed by immediate chilling on ice.
3. Carry out the labeling reaction in a volume of 50 μL including 5× buffer, unlabeled dNTPs (each at 20 μM), denatured template (400 ng), BSA (400 μg/mL), either Cy5- or Cy3-labeled dUTP (2.5 nmol), and Klenow DNA polymerase (100 U/mL) at room temperature for 4 h to overnight.
4. Terminate the reaction by incubation at 95°C for 2 min. Chill on ice and add 20 m*M* EDTA.

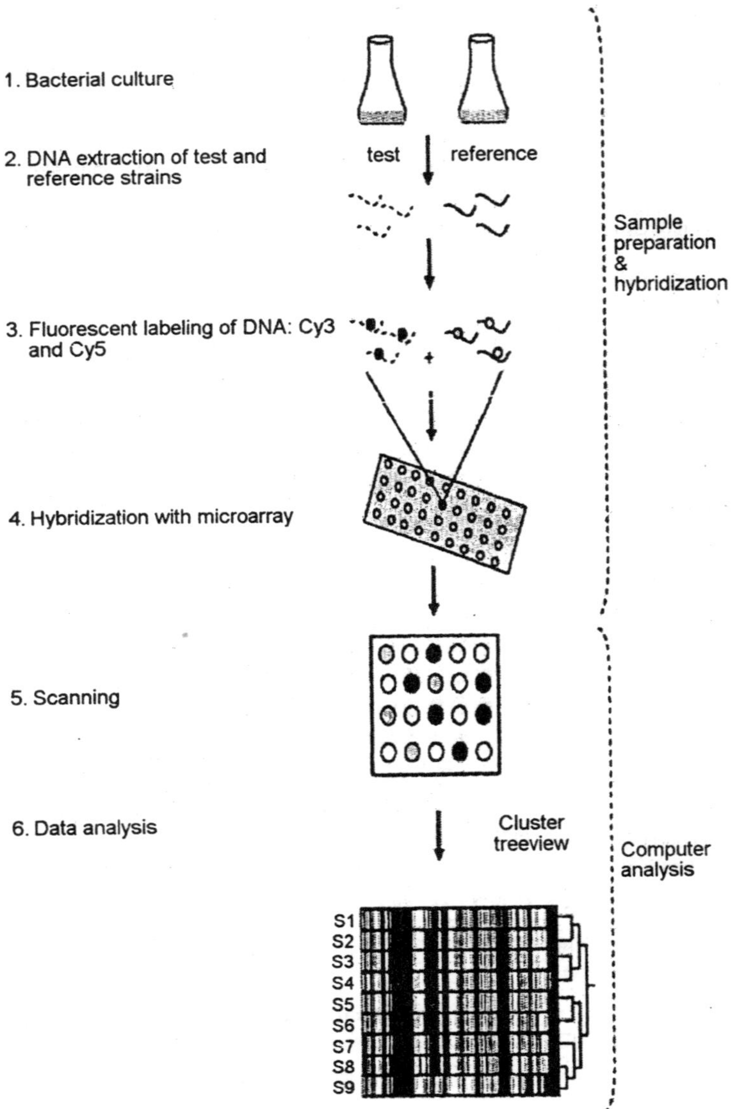

Fig. 9.1. Overview of the protocol for microarray-based comparative genomic hybridization. S1-S9 represent genomic profiles of the strains analyzed

5. Purify labeled genomic DNA using the Centri-spin 20 kit according to the manufacturer's instructions and use immediately in hybridization experiments or store at –20°C for later use.

6. Determine the efficiency of labeling with an Agilent Bioanalyser or equivalent.

Slide Preparation and Pre-Hybridization

1. Add 1.8 mL of hybridization buffer (pre-warmed to 68°C) to the slide and incubate at 42°C for 45 min in a hybridization chamber.
2. Wash the slide in pre-warmed (42°C) distilled H_2O (dH_2O) for 30–45 s and follow by washing in dH_2O at room temperature.
3. Centrifuge the slide at 15 × *g* for 7 min. The slide is now ready for hybridization.

Hybridization and Washing

1. Add equal volumes of dH_2O and hybridization buffer (100 μL each) to the end reservoirs of the hybridization chamber.
2. Mix the labeled probes together with an equal amount of pre-warmed hybridization buffer.
3. Denature the mixture at 100°C for 5 min and add to the edge of the lifterslip.
4. Perform hybridization at 48°C for a minimum of 2 h and as long as overnight.
5. Following hybridization, immerse the slide in 2× SSC, 0.1% SDS (pre-warmed to 42°C) to remove the coverslip.
6. Wash the slide twice in 0.1× SSC containing 0.1 % SDS for 10 min at room temperature, followed by two washes in 0.1× SSC for 5 min and two washes in 0.05× SSC for 2 min.
7. Rinse the slide in dH_2O, followed by 95% ethanol. Dry the slide by centrifugation at 15 × *g* for 7 min.

Scanning of Slides and Data Analysis

1. Scan the slides using a microarray scanner such as a GenePix 4100A instrument or equivalent with line scans performed to maximize signal and minimize background. During scanning, normalize fluorescence intensity between channels to a serial dilution of reference strain chromosomal DNA by adjusting laser power and/or photomultiplier gain. Problems that have occurred throughout the microarray experiment can become apparent upon scanning of the slides.
2. GenePix Pro 6.0 software or similar is used to convert the pixel images into signal intensity before exporting as a flat text file to GeneSpring. For each slide, adjust spot intensities for background and normalize to the median intensity of the slide. The median

ratio of normalized test strain fluorescence to normalized control strain fluorescence is used to identify gene differences. Genes are typically considered lacking or highly variable in the test vs. control strain below a 0.5 test/control ratio and considered present in greater copy number in the test vs. control strain above a 2.0 test/control ratio.

3. Use GeneSpring software for further statistical analysis of the data.

Phylogenetic Analysis

1. To determine the evolutionary relatedness of strains, use the Cluster program based on the presence or absence of all open reading frames (ORFs) represented on the microarray.
2. The output can be represented in a phylogenetic tree constructed using the program TreeView.

Notes

1. ORFs defined as "absent" either are not present in the genome that is being tested or have sufficient sequence divergence that hybridization will not occur with the ORFs represented on the microarray under conditions of high stringency. For each strain, two or three hybridizations are carried out.
2. The reference strain is representative of one of the sequenced strains encoded on the microarray, and the test strain represents a strain of unknown genomic content.
3. To validate the microarray hybridization process, use an array validation system such as the Spot report oligo array validation system.
4. The unlabeled dNTPs are prepared as outlined in the Promega protocol. Briefly, mix 1 μL of each non-isotopically labeled dNTP to yield 3 μL of a premix containing the three unlabeled dNTPs, each at 500 μM. The volume of aqueous labeled dNTPs should not exceed 50% of the total reaction volume.
5. As suggested in the Promega protocol, unincorporated, labeled nucleotides may be removed by size-exclusion chromatography using Sephadex G-50 spin columns or by selective precipitation of the labeled DNA. This step is unnecessary if dye incorporation is above 60%.
6. Streaks of labeled product from strong spots ("comet tails") may be observed after scanning. Steps can be taken to reduce this effect including denaturing of the slides at 95°C for 2 min before

hybridization, reducing the amount of label used in the reaction, or diluting the PCR products before printing the microarray.

7. To minimize high background fluorescence, one may increase the spacing between the spots and lower the photomultiplier tube gain (PMT) during the scanning process. Only genomic DNA of >1.8 $A_{260/280}$ ratio should be used for labeling and hybridization.
8. The occurrence of intense spots in the background of the slide can be minimized by the use of powder-free gloves and working in a dust-free environment. It is also important to ensure that the probe has been sufficiently denatured before hybridization.
9. A global normalization of all samples in addition to individual chip normal-ization should be performed. The reference strain should display a 45° scatter plot whereas the test strain should display a skewed histogram toward the negative. GeneSpring can be used to construct a dendrogram and run a Pearson correlation to determine whether the replicate hybridization experiments are in agreement.
10. The Pearson correlation coefficient is used to represent the scale of a dendrogram constructed with TreeView after phylogenetic analysis with Cluster. For a pairwise comparison, a coefficient of 1 indicates absolute identity and zero indicates complete independence.

10

LABORATORY TECHNIQUES

In nature, almost all kinds of bacteria or fungi, including the disease-producing species, live naturally in more or less intimate association with other kinds. Therefore, the primary culture from almost any source will be a *mixed culture* containing organisms of several different kinds. But in the laboratory the various species may be separated from one another and cultivated by themselves. A culture which contains just one species is called *a pure culture*. An organism must be obtained and studied in pure cultures before the properties peculiar to itself may be learned. The process of obtaining a pure culture by separating one kind of bacterium or fungus from a mixture with other kinds is spoken of as *isolation* of the organism.

PRINCIPLE OF ISOLATION METHODS

The earliest workers to study bacteria in the laboratory used only liquid culture media. The task of securing pure cultures using liquids only was naturally very difficult, and only men of the highest mental attainment and possessed of the patience of genius, like Pasteur and Lister, could make much headway in the study of micro-organisms. We have already explained how *Robert Koch*, in 1881, by the use of solid culture media and the plate method, simplified the entire procedure of securing pure cultures. The methods for isolating bacteria used today are essentially the same as those of Koeb. These methods depend upon (1) *dilution of the original material in solid media cultures in order to secure separated, pure colonies*, and (2) *transfer of a pure colony to a fresh sterile medium*, without contamination from the neighbouring growth. The process of picking out a single colony for a smear or transfer is called *fishing*.

DILUTION METHODS WITH AGAR MEDIA

Just as in planting a vegetable garden it is necessary to scatter the seed over the ground so that the plants which later develop *will* not be too close together, so the bacteria which are to be planted upon culture media must be thinned out so that separate, typical, pure colonies will appear. Any method of diluting the original material will serve as long as extraneous bacteria are not introduced.

Pour-plate Technique

The forerunner of the present pour-plate method was developed in the laboratory of the famous bacteriologist *Robert Koch.* Today this technique consists of cooling melted agar-containing medium (1.5% agar) to approximately 42° to 45° C, inoculating the medium with a specimen and immediately pouring it into a sterile Petri plate, allowing the freshly poured medium to solidify, and incubating the preparation at the desired temperature. In several instances, in which the magnitude of the bacterial population in the specimen to be studied is not known beforehand, suitable dilutions must be created in order to ensure isolated colonies. By means of the pourplate method, bacteria are distributed throughout the agar and are trapped in position. Although the solidified medium restricts bacterial movement from one area to another, it is soft enough to permit growth, which occurs both on the surface and in the depths of the inoculated medium. In addition to the qualitative application of this procedure to isolate and detect bacterial species contained within a mixed culture, it is also useful in quantitative measurements of bacterial growth.

Unfortunately, however, there are several disadvantages to this technique. Colonies of several species may present a similar appearance

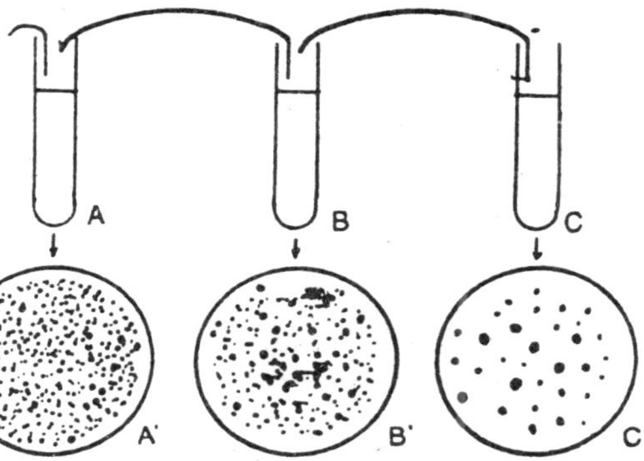

Fig. 10.1. Pour plate culture, illustrating the effect of dilution of the inoculated material.

in the agar environment, thus making differentiation difficult. Certain species of bacteria may not grow under the cultural conditions of the method and difficulty may be encountered in removing colonies for further study.

Streak-plate Technique

This procedure' is also a dilution method. It was originally developed by two bacteriologists, Loefler and *Gaffky*, in the laboratory of *Robert Koch*. The modern method for the preparation of a streak-plate involves the spreading of a single loopful of material containing microorganisms over the surface of a solidified agar medium. The "clock-plate technique" is one of the more generally used forms of this procedure. Here the inoculum is first spread over a small portion of the medium's surface; then the inoculating loop is flamed to destroy any residual bacteria, and the plate containing the medium is rotated approximately one quarter of a turn. The flamed loop is next used to make a second set of streaks, thus diluting (spreading) the bacterial population in the original set of streaks. The original surface streaks are crossed only once. Further dilutions of the specimen are carried out by repeating this sequence; flaming the loop, rotating the medium, and making additional streaks. If this technique is properly performed, well-isolated colonies should grow after incubation at an appropriate temperature. Supposedly, one colony develops from a single cell, thereby producing a pure culture. It is customary to "pick" (transfer) a small portion of a desired colony to a tube of medium, such as broth or agar slant. and utilize the culture as a source of organisms for additional studies.

Another technique, the "spread-plate" procedure, is used in certain types of investigations. Here a bacterial specimen is placed on an

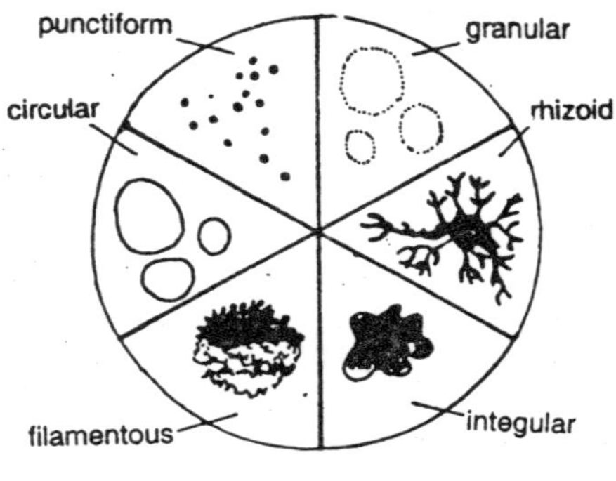

Fig. 10.2. Diagrammatic representation of colonial forms of growth.

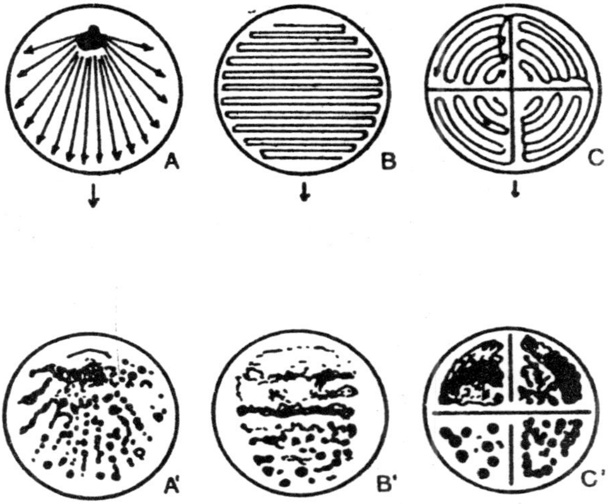

Fig. 10.3. Streak plate culture.

agar medium and spread over its surface with the aid of a sterile bent glass rod. The agar plate can be placed on a rotating wheel device to aid the spreading out of the bacterial specimen. Because of the high concentration of water in agar preparations, condensation forms in most Petri plates. To avoid such *water of condensation* on media surfaces, which may cause bacterial colonies to run together, plates should be incubated in an inverted position.

Fishing

The process of picking up bacteria from a single colony with a sterile needle, without touching neighbouring colonies, is often a delicate task requiring much care and skill. This is particularly true when pathogenic organisms are being isolated because their colonies are usually small.

A straight needle is always used for fishing. Often it is an advantage to have the end of the needle bent at a right angle and filed to a fine point. Colonies to be fished from plate cultures may be marked by drawing a ring about them with a wax pencil on the bottom of the dish. The colonies may be given numbers, and smears or cultures trade from them may be correspondingly numbered. In fishing, as in any other procedure, the culture must be protected from contamination, and the complete removal of the cover of a Petri dish is avoided whenever possible. Fishing from 'plate cultures is much easier if performed with the aid of a low power dissecting microscope.

Study and Identification of Pure Cultures

Once a pure culture is obtained, a long study is usually required before the organism can be fully identified and named. It is necessary to determine: (1) its morphological properties and staining reactions, (2) its cultural characteristics, (3) its physiological requirements and chemical activities, and (4) its pathogenicity (disease-producing power). Also tests would be made (5) to see whether the organism will agglutinate, or otherwise react specifically, with an immune serum known to contain antibodies for the species to which the organism appears to belong. This will identify the unknown organism positively in most cases.

Morphological Study : Staining reactions

Microscopic studies will reveal the form (coccus, bacillus, or spirillum), and the characteristic arrangement and grouping of the cells, their size, and the presence or absence of spores, capsules, granules, or other peculiarities of structure. Examination of hanging drop preparations will show whether the organisms are motile or not. If so, the flagella may be stained. The reaction to the Gram stain is of great help in determining the class to which the unknown bacterium belongs. In case the organism is acid-fast, this will be revealed by the acid-fast staining method. These studies supply a great deal of information about an organism and may give a strong hint as to its identity, but *a species of bacteria can never be identified with certainty on the basis of morphology and staining alone.* There are only three basic shapes which bacteria may take, and only a few other morphological features which help to differentiate organisms from one another, yet there are hundreds of different species. Organisms which stain alike and look alike under the microscope may be totally different in other and more important characteristics. Thus, nearly every type of disease germ is duplicated, so far as morphology and staining are concerned, by other species which are harmless.

Cultural Characteristics

Further information about an organism is obtained by observing the way it grows in various kinds of culture media. Certain species have a very characteristic manner of growth in broth or milk, in agar, or other media. Illustrates some of the forms of growth commonly seen.

The agar *colonies* are sometimes so distinctive that they give a valuable clue to the identity of the organism. Plate cultures are most conveniently used for the study of colony characteristics. The colonies may be observed with a hand lens, or they may be studied under a dissection microscope or under the low power objective of the ordinary microscope. In the description of colonies, the following points are usually considered :*form*, *size*, *colour*, *optical characteristics*, *surface*, *elevation*, *edge*, *internal structure*, and *consistency* when touched with a needle. Thus, a colony of staphylococcus might be described as follows: a circular, orange-coloured, opaque mass, about 2mm, in diameter, with a smooth, glistening surface and regular edge, slightly convex, showing a homogeneous, granular, internal structure, and having a soft consistency.

Physiological Requirements and Chemical Activities

The most significant properties of an organism are learned by a study of its physiology and its biochemical reactions. The following points are among those usually investigated

1. *Food requirements*. Does the organism grow upon the simpler agar and broth media, or does it require media enriched with blood or other body fluids ?
2. *Temperature relations*. Does it grow best at body temperature? If not, what is its optimum growth temperature ? Minimum and maximum growth temperature ?
3. *Relation to oxygen*. Does it grow best under aerobic or anaerobic conditions, or under partially reduced O_2 tension ?
4. *Pigment production*. Does it develop a coloured growth on agar slants or other cultures?
5. *Proteolytic action*. Does it liquefy gelatin? Cause digestion of coagulated blood-serum, meat, and similar substances ? Does it form indol ? (Indol is a product of the decomposition of tryptophane, one of the amino-acids composing proteins.)
7. *Fermentation of carbohydrates*. Does the organism produce acid, or acid and gas, in culture media containing dextrose, lactose, maltose, sacchrose, or other carbohy-drates?

When these facts are known it often becomes possible to recognize the species, or at least the group of closely related species, to which the unknown organism belongs.

PATHOGENICITY

One other important property remains to be tested, however Will the organism produce disease in laboratory animals ? The animals most commonly used in laboratories are guinea pigs, rabbits, white mice, and white rats. For certain special work monkeys, dogs, cats, or other animals may be needed. Pathogenicity tests are usually made by inoculating the organisms under investigation into these animals *subcutaneously* (*just under the skin*), *intraperitoneally* (into the peritoneal cavity), or *intravenously* (into a vein). Other possible routes of injection are *intradermal* (into the skin just under the epidermis), *intramuscular* (into the muscles, as of the thigh), and *intracerebral* (*directly* into the brain through the skull).

Such animal inoculations are often of great value in identifying a particular organism. Also they help in securing pure cultures of disease germs. If a material contains acid-fast bacilli, for example, and there is doubt as to whether they are tuberculosis bacilli or not, the matter can be settled by inoculating a guinea pig. If the material injected contains tubercle bacilli, the guinea pigs will develop tuberculosis in a few weeks. It is difficult to cultivate the tubercle germs upon culture media directly from sputum or other matter containing them, but if this material be inoculated into a guinea pig, the germs can easily be obtained in pure culture from the lesions in the animal.

Identification by Agglutination Reactions and Other Serological Tests

One final method of identifying a species of bacteria is by use of serological tests. If, for example. we have a Grain negative bacillus which has the characteristics of the typhoid gear. a final test of its identity may be made by mixing a suspension of the organisms with a blood-serum known to contain antibodies for that species. If agglutination occurs under properly controlled conditions, the organism must be a typhoid bacillus.

Counting Bacteria by Cultural methods

In the examination of water, sewage, milk and other materials, it is often necessary to make a count of the number of bacteria. The method is simple, but the cultures must be made very carefully in order to assure consistent results. The *standard methods* for counting the bacteria in water and milk, published by the American Public Health Association, are used almost universally. It is necessary (1) to secure a representative sample of the material to be examined in a sterile container. Then (2) the sample is diluted in a definite quantitive manner and (3) the diluted material is inoculated into agar media. Then (4) the colonies which develop are counted, and, assuming that each colony represents a single living organism in the original sample,

(5) the number of bacteria per cubic centimeter of the sample is calculated.

Technique

The method of counting of bacteria in water may be taken as an example. The technique is as follows:

1. Shake the water sample thoroughly, and transfer, with a sterile pipette, lc.c. of the sample to a bottle containing 99 c.c of *sterile* water. (This is called *a water blank*). The sample is thus diluted 1:100. If 1 c.c of this thoroughly mixed dilution is added to another 99 c.c . water blank, this will give a dilution of 1:1000. As many dilution as are necessary (depending upon the number of living organisms in the original sample) are made in this way.
2. Place 1 c.c. of each dilution in the bottom of a separate sterile Petri dish. Mark the dishes accordingly. Then pour melted agar, cooled to about 45°C, into each dish, and mix it thoroughly with the water by gently tilting and rotating the dish. It is best to make these plates in duplicate or triplicate.
3. Allow the agar to harden. Then invert the dishes and incubate them for twenty-four to forty-eight hours.
4. Discard those plates which contain more than 300 colonies, because a count based upon them could not be accurate. Count the colonies in each of the other plates. Each colony is assumed to represent a single organism in the original water sample. Multiply this individual plate count in each case by the dilution of the sample in that plate, so as to get the number in 1 c.c. of the original sample. Take the average of the individual plate counts as the final figure.

Condition of Incubation

Aerobic Incubation

Microbial growth, as noted previously, can be dependent upon suitable levels of oxygen and carbon dioxide. While aerobic organisms may be able to grow under anaerobic conditions, the obligate aerobes, such as *Acetobacteri* spp., must be in intimate contact with air, or approximately 20 per cent oxygen. Many organisms have less stringent oxygen requirements and will grow well in the depths of liquid or solid media. Therefore, cultivation of aerobic or facultative anaerobic bacteria is relatively simple under usual laboratory conditions.

Once an appropriate solid or liquid medium has been inoculated and protected from contamination, all that remains to be controlled is the,. temperature and occasionally the level of humidity. Standard laboratory incubators usually are sufficient to maintain such environmental factors.

With any incubator, it is advisable to employ a maximum minimum thermometer to determine the temperature range provided by the thermostatic control. One type of maximum minimum recording thermometer consists of a bimetallic coil (in the form of a strip), a dial equipped with a needle indicator to register the temperature, and two hands. The hands are moved by the temperature indicator ; one remains set at the highest and one at the lowest temperature value. Periodic recordings of the daily extremes give advance notice of the possibility that the controls may be wearing out, and can provide warning of the instrument's malfunctioning in time to repair it before any serious damage occurs, such as cooking the cultures.

Capneic Incubation

Air enriched with carbon dioxide is required by some bacteria, including *Neisseriagonorrhoeae*, *N. meningitides*, and *Brucella spp.*, for their primary isolation from clinical specimens. As mentioned earlier, many other *microorganisms*, such *as Mycobacterium tuberculosis*, appear to grow better if allowed to incubate under these capneic conditions. This form of enriched atmosphere can be supplied in a variety of ways. The candle jar technique is routinely used for *N. gonorrhoeae* cultivation. With this technique, Petri plates and/or tubes with inoculated media are placed in a jar with a candle. The lighted candle will continue to burn after the lid has been placed on the jar,

Fig. 10.4. The candle jar.

until the CO_2 concentration increases to the point which stops combustion, usually around 3 to 5 per cent CO_2. This should not be considered to represent complete combustion of oxygen as aerobic organisms will grow and strict anaerobes will not grow in such a candle jar.

For a laboratory which is primarily concerned with the maintenance of *N. gonorrhoeae* on slanted media, the test tube capneic system is worthy of consideration. In this system a capsule of sodium bicarbonate is added to a small tube of dilute sulfuric acid and the large tube is immediately stoppered. The reaction here is the same as that used in fire extinguishers: acidification of bicarbonate to release CO_2.

Anaerobic Incubation

Certain anaerobic microorganisms are known to be responsible for a variety of diseases. while others are functional in several industrial fermentation processes. Some of these anaerobes appear to be - *aerotolerant* they can withstand oxygen to a limited degree. Others of them can survive only in an environment without oxygen and *are obligate anaerobes.* One explanation for their sensitivity to an oxygen environment is related to the role of the oxygen in the formation of hydrogen peroxide (H_2O_2) during microbial metabolic processes. This product alone can be quite destructive, as shown by the fact that it is commonly used as a disinfectant. Some bacteria produce an enzyme. catalase, which breaks down the peroxide to oxygen (O_2) and water (H_2O). Strict anaerobic bacteria do not appear to have catalase. and are therefore poisoned by the peroxide.

An additional explanation for the lack of growth by anaerobes in the presence of oxygen concerns the oxidation-reduction potential (Elm) of the media, which may be the more significant mechanism responsible for the anaerobes oxygen intolerance. The oxidation-reduction potential is an electrical phenomenon that indicates the electron-accepting the yielding potentialities of substances. When the Eli of a medium is lowered by the addition of a reducing or oxygen-absorbing compound, such as cysteine or thioglycollic acid, many anaerobes will grow.

Anaerobic Transfer

The lack of growth of many clinical isolates may not be due to the Eh of the properly prepared anaerobic medium, but in reality to transient exposure to air during the piOcedure for specimen collection, transportation to the laboratory, or after groom, during the transfer of organisms to subculture media. In order to prevent inactivation of anaerobes when transferring them, the operation should be performed

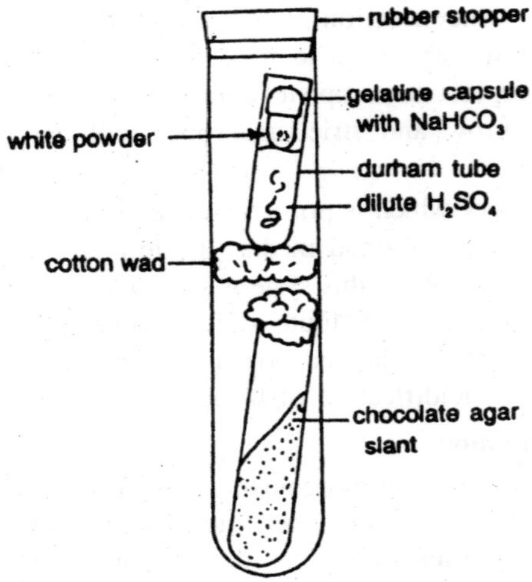

Fig. 10.5. The test tube capneie system.

mouse of a pipette stream of nitrogen for cultures on solid media, or by the use of a pipette with liquid media. For example in the transfer of cultures from thioglycollate broth to a fresh tube of medium, a pipette should be introduced into the bottom of the tube with the finger on the mouth-piece. By carefully lifting the finger. inoculumn from the most anaerobic area is allowed into the pipette. Next, with the finger on the moullipiece, the pipette is removed and inserted into the bottom of a fresh tube of thioglycolate. The finger than is quickly removed, which causes the release of the inoculum into the most anaerobic portion of the medium. Following this step, the pipette is removed and the newly inoculated preparation is incubated at the desired temperature.

Methods of anaerobic incubation

Incubation under anaerobic conditions can be accomplished with vacuum devices. Incubators for this use are usually fitted with gaskets and Ports for evacuation and flushing with inert gases. One method commonly used involves flushing a specially equipped incubator with 95 per cent N_2 and 5 per cent CO_2 to remove any oxygen and to introduce CO_2 into the system. The latter gas may enhance growth. Hydrogen also may be used for flushing, but it forms an explosive mixture with oxygen. If for some reason anaerobic hydrogen bacteria are to be grown hydrogen should be bubbled through water first; an explosion will not occur with wet hydrogen.

Before the advent of the anaerobic incubator, *Brewer* invented two systems for culturing anaerobic bacteria. In one system thioglycollate agar is inoculated and poured into a Petri dish. The cover for this Brewer plate forms a seal with the surface of the agar and with the edge of a slight concave opening, to allow a limited surface for organisms on which to grow. The trapped oxygen is rapidly reduced by the thioglycollate, thus producing anaerobic conditions.

An alternative to the Brewer plate is the Spray dish. This device consists of a deep, double-well dish and cover. The intent is to permit separation of an alkaline solution, such as potassium hydroxide (KOH), and pyrogalic acid. These chemicals when combined absorb oxygen and produce anaerobic conditions. When this system is used, an inoculated Petri plate is inverted over the bottom dish and sealed into place. The dish is then tipped to permit the mixing of the chemicals, thus resulting in anaerobiosis.

The Brewer jar was developed in an attempt to have a larger incubation system for anaerobic organisms. In this system, the inoculated plates and/or tubes are placed in the jar, the cover issealed into place, the air in the container is immediately evacuated, and the system is

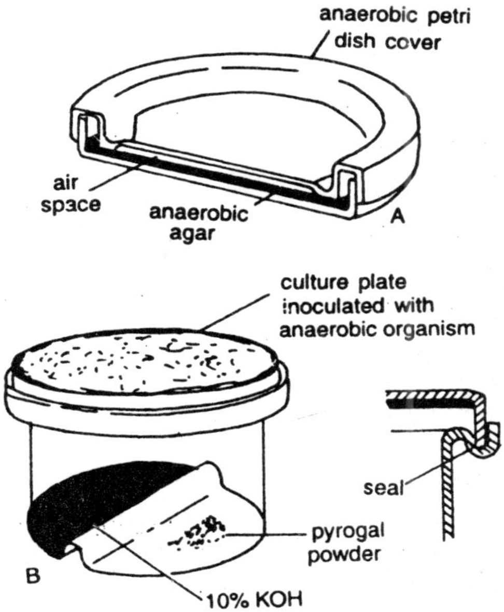

Fig. 10.6. A cut-away drawing of a Brewer Petri dish. B-spray dish for anaerobic incubation.

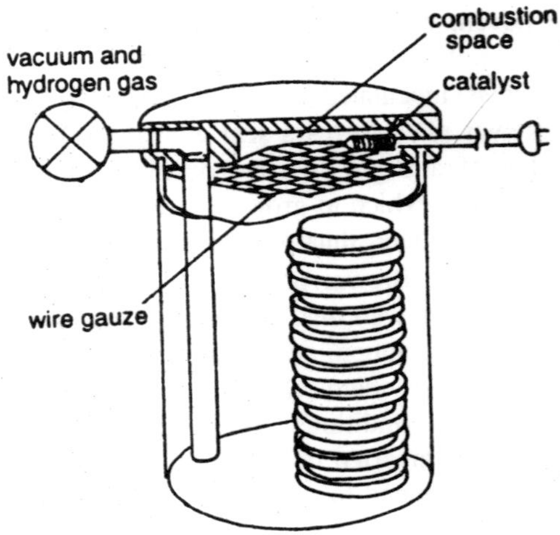

Fig. 10.7. The Breiter anaerohic jar.

flushed with hydrogen. Subsequently, a catalyst, such as platinum, is heated in the lid by an electric current, with the result that the hydrogen combines with residual oxygen to form water. A slight vacuum is created in the chamber due to the combustion reaction.

A recently developed alternative to the Brewer Jar is the Bio Quest disposable Gas Pak. This enveloped unit is a self contained hydrogen generating and catalyst system. much simpler to use than the Brewer Jar. The plates and or tubes are placed in ajar with a Gas Pak envelope. The envelope is opened, water is added to the reagents, and the jar is sealed immediately. The generation of hydrogen in the presence of air and the catalyst causes reduction of the oxygen to water, which condenses on the side of the jar. This system eliminates the need for hydrogen and nitrogen tanks and vacuum pumps, which greatly enhances the ability of a small clinical laboratory to carry out anaerobic cultivation.

Anaerobic Indicators and Checking the Efficiency of Anaerobic Systems

Anaerobic incubation systems should be checked whenever possible to provide assurance that adequate conditions have been obtained and maintained during incubation. One system used for this purpose incorporates Bacto-OR Indicator Agar (Difco Laboratories). This medium uses methylene blue as an indicator of oxidation-reduction potential. It is usually dispensed into small screw-capped vials and autoclaved (sterilized under pressurized steam). This reduces the methylene blue to a colourless state, leucomethylene blue. The procedure for testing involves placing an opened vial of medium in the incubator and initiating the anaerobiosis environment. After a suitable incubation period the vial is removed and examined. If anaerobic conditions were maintained during the test period, the methylene blue should still be in the reduced state, i.e., leuco. However, if anaerobiosis was not achieved, the medium will be oxidized and appear blue. (Note that, in the first case, the brief exposure to air may cause a slight oxidation of the leuco- methylene blue at the surface of the medium).

Another system for checking the efficiency of an anaerobic incubator involves subculturing an obligate anaerobic bacterium in the same incubator as other specimens. If the indicator organism grows, satisfactory conditions were met. Organisms that may be used for this purpose include certain *Clostridium* and *Bacteroides* spp. However, the use of chemical indicators is recommended due to their simplicity and reliability.

Biological or "Sterlity Test" Cabinets

A definite need exists to provide protection from airborne contamination. Such units have value in laboratories that handle "biohazardous" substances, the preparation bacteriological media, sterility testing, and the inoculation of a variety of media.

Cabinets of this type are equipped with a double polycarbonate viewscreen which provides a functional visual working angle and contributes to the protection of individuals within the confines of the working area. A recirculating vertical airflow unit creates a "front air barrier" at the opening of the work area. This "barrier" protects both the laboratory personnel and materials from contamination by preventing airborne particulates from leaving or entering the cabinets. Air entering and leaving the working area passes through HEPA (High Efficiency Particulate Air) filters which remove particles of 0.3 μm (μ) and greater. Although contamination of personnel and materials is prevented in a unit such as the one described, workers must still observe aspetic precautions and techniques.

Culture and Susceptibility

Many microbiologists believe that the culture and susceptibility is the most important clinical test performed, because the results will provide the physician with significant information about the pathogen and how to control it. *A culture and susceptibility test* is the growth of a known pathogen in pure culture and its exposure to numerous antibiotics of different concentrations to determine which drug will kill or inhibit growth of the microbe. The smallest amount of the drug that prevents the microbe's growth is called the *minimal inhibitory concentration (MIC)*. The drug with the lowest MIC is usually the one selected by the physician for use in treatment. The culture and susceptibility test, and determination of the MIC have become greatly important due to the increasing number of antimicrobial agents available to physicians, and because of an increase in the number and types of pathogens that have become resistant (unaffected) to these drugs. Two methods commonly carried out to determine the susceptibility of a microbe to an antimicrobial agent are the tube dilution method and the agar diffusion (disk susceptibility) method.

The *tube dilution method tests* the susceptibility of a bacterium to antibiotics and is performed by inoculating the bacterium into a serial dilution of the antibiotic. Each tube contains an appropriate culture medium and a specific concentration of the anitbiotic. If the bacterium is inhibited by the drug, there will be no growth in the tube and the medium will remain clear. If microbe is resistant to the effects of the drug, it will grow and produce a cloudly tube. Several antibiotics are compared using this method in order to determine which inhibits the bacterium in the lowest concentration. If more than one drug successfully controls growth, all the inhibiting drug's names are sent to the physician for consideration. This testing procedure may be used to determine the susceptibility of microbes from blood culutures, the susceptibility of patients who have undergone a relapse during therapy, and the susceptibility of patients who have not responded to the use of an antimicrobial drug. This method is time-consuming and becomes fairly expensive when testing a number of antibiotics. Therefore, most labs are prepared to run a tube dilution series, but prefer the agar diffusion (disk susceptibility) method.

The *disk susceptibility* (sensitivity) method is probably the most widely used and quickest susceptibility test used in hospital labs. In this method, the medium (usually Mueller-Hinton) is inoculated by swabbing the bacteria over the entire surface so that they will grow into a "lawn." Small disks of paper impregnated with various antibiotics of different concentrations are placed on the surface of the plate. The plate is incubated overnight and checked the following day for growth inhibition. During the incubation period, the antibiotic molecules diffuse from the disks over the surface of the bacteria-coated agar. If the concentration of an antibiotic is strong enough to inhibit the bacteria's growth, there will be a zone of inhibition around the disk impregnated with that drug. If the bacteria are resistant to the drug, there will be little or no zone of cleaning and the bacteria grow close to the disk. The diameter of the inhibition zone designates the relative effectiveness of the drug or the bacteria's susceptibility to the antibiotic. The zone is measured from the edge of the antibiotic disk to the edge of the cleared area that shows no visible growth. The size is then compared to a table that has been developed from disk susceptibility tests run on known *EE coli* and *Staphylococcus aureus* culture. This is known as the Kirby Bauer technique. The table lists the antibiotic agents, concentrations used in the disks, and the inhibition zone diameters measured from the standard E. *coli* and *S. aureus*. The inhibition zones are divided into three ranges; resistant, intermediate, and susceptible. Disk susceptibles that show clearing only to the resistant level have not been inhibited by the drug. Drugs that have cleared to the intermediate range may be used to control the pathogen, but may be required in high dosages. If this is the case, another drug may better be used. Dmgs that have resulted in large zones of clearing on the agar surface and fall in the susceptible range are those that may be expected to inhibit the pathogen in the patient. They may be recommended for use in controlling an infection caused by the bacterium being tested. The disk susceptibility method is quick and accurate, but there are several sources of error in the test. These possible errors make it essential that a qualified technician perform and interpret the results.

1. Inoculum density not carefully standardized larger zones of inhibition may result with a light inoculum, smaller zones with a more dense inoculum.
2. Using a mixed culture: the standardized test is based on the use of pure cultures, a mixed culture can yield completely erroneous susceptibility results.
3. Excessive moisture on the surface of the medium may yield very confluent growth and reduce the diameter of the inhibition zone even though the density of inoculum is correct.
4. A very dry surface may result in poor growth, yielding a larger inhibition zone.
5. Mishandling of susceptibility disks by permitting moisture uptake and prolonged exposure to room temperature results in deterioration of most antibiotics.

All of the information gained from culturing microbes is sent to the physician. The physician then interprets this information and decides on an appropriate course of treatment in light of the patient's condition (both mental and physical), the available health care facilities, chemotherapeutic drugs, and the time needed for treatment. In order to ensure successful treatment of the patient and containment of the infection, the lab must perform another essential service known as *sterility testing.*

Hospitals use a great many sterile materials in the care and treatment of patients. Bandages, dressings, hypodermic needles and syringes, glassware, scissors, catheters, and surgical instruments must all be checked for contamination by bacteria, fungi, and yeasts before they are used. Usually a sample from a batch of sterilized instrument is taken to the lab for testing. If the sample is found to be free of all living things, the entire batch is considered sterile and safe for use. The item to be sampled is opened using proper aseptic technique and sampled either with a sterile moist swab or with RODAC plate. (RODAC is an acronym for Replicate Organism Detecting and Counting). Swabs are typically used to sample hard surfaces such as utensils, glassware, or instruments while the RODAC plates are used to sample linens, gauze, and flat surfaces.

The RODAC plate is a small petri dish filled with so much culture media that it can be pressed directly onto the surface of the material being tested. The plate is incubated for forth-eight hours after which the colonies may be counted and identified. Another technique is used to check the efficient operation of the autoclave and other hospital sterilization equipment. Special paper strips containing endospores of the bacteria *Bacillus* stearothermophilus and Bacillus subtilis are placed in the autoclave with the material to be sterilized...,

Collection of Specimens for Bacteriological Examination

The laboratory examinations of pus, urine, feces, sputum, or other material collected from the living patient, is relied upon to furnish the final diagnosis, and often to determine or control the treatment, of many of the most common diseases. A bacteriological study of material secured from the dead body at autopsy often serves to explain the cause of death.

The value of the laboratory examination of any material is in large part determined by the care and intelligence with which the specimen is collected. If the proper technique is not employed at the time of collection, or if the specimen is not properly cared for after collection, an examination of it may be impossible or useless. It is frequently desirable to reach a diagnosis in as brief a time as possible in order to permit the proper treatment to be given, and delay or error in diagnosis, by reason of an improperly collected specimen, may be a very serious matter.

General Rules

Secure an uncontaminated specimen

Avoid contamination of the specimen with organisms from other sources. Use *a sterile container*, and protect the specimen from contamination with dust bacteria as you would any culture tube. When cultures or smears are made directly from an infected part of the body, *sterile swabs* or *needles* must be used, and the greatest care must be taken to avoid picking up organisms from adjoining tissues or other sources. This is not in every case an easy matter, but an effort must always be made to secure as nearly pure a specimen as possible.

Label specimen fully and accurately

Every specimen, smear, or culture must be fully, clearly, and accurately labelled. The label should indicate unmistakably the patient from whom the material was obtained, the date, the nature and source of the specimen, and the kind of examination desired. Special pains should be taken to see that the name attached to a specimen is the proper one, for serious results may follow failure to associate a patient with his own specimen.

If a printed form is to accompany the specimen to the laboratory, this must be filled out completely, accurately, and in a legible manner. The bacteriologist is able to save time, to perform a more intelligent examination, and to preserve more accurate records if the fullest possible information is given him as to the clinical state of the patient, the exact nature of the specimen. the time or the manner in which it was secured. and the kind of studies he is expected to carry out.

Do not add a disinfectant unless this is especially ordered

As a rule disinfectants are not to be added to a specimen intended for bacteriological examination, because the bacteriologist may wish to cultivate the microbes in the specimen or to observe them in a living state. The only exception to this rule is in the case of sputum or feces from a case of suspected tuberculosis. Ordinarily, this material is used solely for making smears. But here also it is best to omit the disinfectant, so that the bacteriologist may make cultures from the specimen if he desires.

Handle with care to avoid infection

Since no disinfectant is added, specimens may contain very many living germs in a highly dangerous state. The outside of the specimen bottle or other container should not be soiled, so that it may be handled without contaminating the fingers.

See that specimens reach the laboratory promptly

Place cultures in a 37.5°C incubator, and other specimens in a refrigerator, immediately after collection. It is often very important that the examination be made as soon as possible after collection. The best results are obtained when the bacteriologist brings his swabs and culture media to the bedside, or takes his specimen from the patient directly to the laboratory. Failing this, the most reliable results are secured when the material is studied within one or two hours after collection. A specimen of pus, feces, or sputum is practically useless for cultures when more than four hours old.

No time muse be lost in getting cultures to the incubator

The incubator temperature of 37.5°C. (body temperature) will favour the development of the pathogenic types of organisms which are being looked for. But if the culture is permitted to stay long at room temperature, some nonpathogenic types may grow so rapidly that they overwhelm the disease germs.

Specimens of urine, feces, sputum, pus, spinal fluid, and similar material must often be kept for a short time before they can be examined. *During this interval between collection and examination, they must be kept continuously in a refrigerator.* Usually a single kind of pathogenic organism is to be looked for in a specimen- the typhoid bacilli in feces, for example, or pneumococci in sputum. If the specimen is allowed to stand in a warm place, the numerous other kinds of bacteria always present in the.tnaterial night multiply to such an extent that the germs would be very difficult to find.

Collection of Particular Specimens

Pus From wounds and other lesions is usually secured with a sterile cotton swab. Abscesses are sometimes lanced and a sample of the pus collected on sterile gauze or in a sterile test tube. A cotton-plugged sterile capillary pipette may often be used to advantage ; its tip may be sealed in the flame before sending it to the laboratory.

Sputum for bacteriological examination should be collected carefully, with attention to a number of details. The mouth should be clean; the patient should rinse it with a mouth wash just prior to collection of the specimen. The specimen must contain true sputum, raised from the deeper air passages by *a cough*, and not merely saliva. The best time to collect the specimen is in the early morning, because patients frequently find it easier to raise sputum at the first expectoration in the morning than at any other time. The sputum must be received in a sterile Petri dish or a clean cup and transferred to a sterile widemouthed container. Sputum is examined in cases of pulmonary tuberculosis, pneumonia, influenza, whooping cough, bronchiectasis and lung abscess, and in other forms or acute infection of the deeper parts of the respiratory tract.

Feces are collected in a clean, preferably sterile pan. Portions may be placed in sterile wide-mouthed bottles or large test tubes, making the transfer with sterile sticks or large swabs. Fecal material may be secured from very young infants by inserting a sterilized glass tube into the rectum. The feces are always examined in cases of typhoid and paratyphoid fever, dysentery, cholera, and other diseases of the intestinal tract. Cultures are often made from the feces of healthy persons in order to discover carriers of typhoid bacilli and other germs. Feces are examined microscopically for hookworms and other intestinal parasites.

Urine is collected for bacteriological examination in cases of suspected infection of the urinary tract. It also may be examined in typhoid fever and other intestinal diseases, in tuberculosis of the urinogenital organs, and occasionally in pneumonia or septicemia, and other diseases. The external genitals and the lower part of the urethra always contain many bacteria, and these must be avoided in order to -secure an uncontaminated specimen of urine. A satisfactory sample may be obtained from males if the glans and meatus are first thoroughly cleansed with soap and water. The first urine voided. should be discarded, and only the last portion collected in a sterile, wide-mouthed bottle. From a female, a urine sample satisfactory for bacteriological examination can be secured only by use of a catheter- a small rubber tube inserted into the bladder or ureter. The part must first be washed with soap and water. The catheter must be sterile, and the first portion of urine should be rejected.

Peritoneal or *pleural fluid* is collected by inserting a sterile needle through the wall of the abdomen or chest and withdrawing the fluid in a sterile syringe or allowing it to run into sterile test tubes. The skin at the site of the puncture must be disinfected, of course, and every aseptic precaution must be taken as in any operation.

In heart and kidney disorders, and in some other organic disease conditions, there may be an accumulation of very large amounts of sterile fluid in the peritoneal cavity. This condition is cared ascites and the liquid is called *ascitic fluid.* Similarly, an excess of fluid may accumulate in the pleural or pericardial cavities. *Sterile fluids* of this sort, originating from some cause other than an infection, are called *transudates. Ascitic fluid*, aseptically collected, is used in the bacteriological laboratory as an enrichment for culture media.

Spinal fluid is obtained for examination by a process known as *lumbar puncture.* A large sterile hollow needle or trochar is passed into the spinal canal through a sterilized and anesthetized spot on the skin of the back, slightly to one side of the midline, between the third and fourth lumbar vertebrae. The strictest asepsis is observed throughout this operation and at its completion the puncture wound is protected with a sterile dressing. In pathological conditions the spinal fluid is usually much increased in amount and under much greater pressure than normal, so that it flows readily from the needle. It may be allowed to run directly into sterile centrifuge tubes. When fluid from the same patient is collected in several tubes, these must be numbered to show in what order the specimens were obtained. Often blood appears in the first drops of fluid collected, while later, unmixed spinal fluid is secured.

The spinal fluid is normally a clear liquid, containing only a few leukocytes, and *sterile. Therefore*, if bacteria, or an unusual number of pus cells, are found in it, this is evidence of a diseased condition. The fluid may still appear clear and yet be very abnormal. In syphilis and infantile paralysis (poliomyelitis), it is usually clear; and in tuberculous meningitis, it is either clear or slightly cloudy. In cases of acute meningitis, however, due to *Sterptococcus*, *Pneumococcus*, *Meningcoccus* or other bacteria, the fluid is usually very cloudy and contains many pus cells.

Spinal fluid from cases of syphilis is not used for cultures or smears. but only for a Wassermann or other diagnostic test. The sample should be sterile.

Blood

Many important facts are learned from examination of the patient's blood. The blood may be collected for study of the blood cells, or for tests on the blood-serum, or for smears or cultures to detect micro-organisms that may be in it. There are two common methods of securing blood for examination. These are: (1) the skin puncture method, by which a few drops of blood are secured from a puncture wound of the finger or ear, and (2) the venous puncture method, by which a considerable amount is obtained through it hollow needle inserted into a vein.

Technique of Skin Puncture Method

The lower lobe of the ear or tip of the finger is prepared for puncture by cleaning throughly with and alcohol sponge. An increased flow of blood to the fingers may be secured by shaking or swinging the hand vigorously, and the ear may be rubbed gently. After the alcohol has dried, the skin is then punctured, *not cut*, with a quick determined movement. A special blood lancet may be used for the puncture, but a sharpened, stiff needle or pin, or half of an ordinary writting pen, will serve. The puncture needle must be sterile. The blood should well out of the puncture wound spontaneously. It is best to avoid squeezing the ear or finger in order to secure enough blood. When the bleeding process is completed, the puncture wound is wiped with an alcohol sponge or touched with tincture of iodine.

Use of Skin Puncture Method

This way of obtaining drops of blood serves for smears to be examined for malaria parasites and other organisms.

Also this method is used to secure blood for a determination of the number of red blood cells ("red count"), or white blood cells (" white count"), and for making smears from which to determine the relative proportion of the various types of white blood cells ("differential count"). The amount of hemoglobin the coagulation time, and other properties of the blood may be tested also.

The cell counts are made by drawing up a small amount of blood from the drop on the finger or ear into special blood counting pipettes, which accurately measure and dilute the blood. Drops of this diluted blood are examined in special chambers under the microscope and an actual count is made of the number of red or white blood cells in a known number of blood. From this, an estimate is made of the total number of the cells in a cubic millimeter of the patient's blood. The differential count is made by counting the first 200 white blood cells seen in a stained smear of the blood and classifying them into the various types- the polymorphonuclear leukocytes, lymphocytes, and so forth. The percentage of each type present is then determined. This differential count and the white count are very important in case of germ disease, because they serve as an index of the severity of the disease and of the resistance of the patients has.

Technique of venous Puncture Method

In adults the veins at the bend of the elbow are used. A bandage of cloth or a tourniquet of soft rubber tubing is applied to the upper arm and the patient is made to clench the fist. This makes the veins at the elbow and lower arm stand out prominently. The skin about the veins is cleansed with soap and water (if necessary), then washed with 70 per cent alcohol. Tincture of iodine may be used. A sterile hypodermic needle is then inserted into the vein and the blood is caught in a sterile tube. When the bleeding is complete, the tourniquet is immediately removed and an alcohol sponge is pressed firmly against the puncture wound until the bleeding stops.

When it is necessary to secure blood from an infant, for the Wassermann or other tests, the skin of the finger or heel may be punctured, or the blood may be secured from the large vein that lies just below the anterior fontanel (the soft spot on top of the head). This area is shaved and cleansed, then a sterile needle is thrust into this vein and the blood withdrawn into a syringe.

Use of venous Puncture Method

This method is used whenever considerable amounts of blood are required for chemical or bacteriological tests upon the *blood serum*. Blood for the *Wassermann test* (for syphilis) is always secured in this way. The blood is allowed to *clot* in the sterile collection tube, and the clear serum which separates off from the clot is used for the test. The specimen should be sterile, and must be kept in a refrigerator until the test is made.

Culture and Smears from the Living Patient

Blood Cultures

The blood of healthy persons is sterile

But in various diseases living bacteria may be present in it and cultures of the blood serve as one of the most important methods of diagnosis. Usually 10 c.c or 15 c.c. are with drawn with a syringe from the arm vein as described above. *The blood is at once expelled into a flask of sterile broth.* Bile broth is used when typhoid bacilli are likely to be present in the blood, and brain broth when the presence of streptococci, pneumocoeci, meningococci, or other organisms is suspected. Some of the blood may be added to melted agar the agar poured into Petri dishes. If the blood must be carried some distance to the laboratory before it can be inoculated into media, it can be caught in a flask containing a sterile, 10 percent solution of sodium citrate. Enough of this solution should be used to give a final concentration of about 0.1 per cent of sodium citrate after mixture with the blood. This will be sufficient to prevent clotting.

Cultures from the nose and throat

Material for nose cultures is secured by passing a small, sterile, cotton swab straight back through the nostril. The swab must not be large, or loose upon the stick, otherwise it may catch in the nose.

Cultures from the throat are made by touching a sterile swab to the *diseased part only*. There must be plenty of light in the throat so that it is possible to see exactly what is being done. When the presence of diphtheria bacilli in the nose or throat is suspected, the medium used must be Loeffler's serum medium. For all other purposes, blood agar is the most suitable medium.

Cultures from the nasopharynx

The nasopharynx is the upper part of the throat back of the soft palate. Cultures from this region are made by use of a special swab. The cotton is wrapped about the end of a wire having a nearly right angle bend, which is enclosed in a glass tube similarly bent. The tube is introduced into the mouth and the curved end passed back of the soft palate. The swab is then pushed out and rubbed across the nasopharynx, then withdrawn into the tube before it is removed from the mouth.

Other cultures

Cultures from the *eyes* may be made by rubbing a small sterile cotton swab wet with sterile salt solution very gently over the surface of the conjunctiva. Cultures from the *cervix* or *uterus* can be made only when the parts are exposed by the use of a sterile speculum. The very greatest care is required not to contaminate such cultures with organisms from the vagina.

Direct smears

There is frequently more to be learned form a smear made directly from an exudate or lesion than from cultures of the same material. Smears alone are relied upon for the diagnosis in cases of acute gonorrhea. Vincent's angina, and in some other diseases. Often the examination of smears from the pus of abscesses, wounds, inflamed eyes or ears, and other lesions, gives important clues as to the cause of the infection. Smears must be *thin*, and of good size. The slide must be clean and not greasy. When smear is to be made with a swab, *roll* the swab over the slide and avoid unnecessary pressure, so that the pus cells will not be broken up.

11

Control of Bacteria

Definition of Terms

Sterile

Sterile means free from life or micro-organism of any kind. This is achieved by sterilization.

Bactericides

These are agents having the property of killing bacteria. The action is irreversible in this case as the organism is killed.

Bacteriostatics

These agents have the property of inhibiting multiplication of bacteria. But on removal from the agent, bacteria resume their usual multiplication. The action in this case is reversible.

Disinfectants

These are agents that free from infection, i.e., destroy or kill infecting organisms and are bactericidal in nature.

Antiseptic

This literally means an agent that removes the cause of sepsis. These are agents which prevent the growth of bacteria and counteract sepsis, putrefaction and decay. They are bacteriostatic in lower concentrations, but kill organisms when used in higher concentrations and act as disinfectant. The terms antiseptic, germicide and disinfectant are practically synonymous.

Asepsis

It is characterized by absence or prevention of the growth of pathogenic microbes. This is achieved by taking sterile precautions as in surgical practice.

The action of disinfectant or an antiseptic depends on : (a) the organisms—vegetative form or spores, (b) the strain—whether it is rough or smooth, (c) the growth phase of the organisms being most sensitive to inimical agents in the log phase: and (d) nature of the bactericidal agent. Bactericidal agents act in different ways by :

(i) Coagulation of the bacterial protoplasm, e.g., heat, heavy metals.

(ii) Disruption of cell membrane. The cell membrane acts as a selective barrier. Substances, which concentrate at the cell surface, may alter the physical and chemical properties of the cell membrane, preventing its normal functions and thereby killing or inhibiting the cell.

(iii) Oxidizing or burning our the protoplasm, e.g., halogens.

(iv) Interfering with bacterial metabolism by way of preventing the intake of nutrition by affecting their enzyme and coenzyme systems.

Disinfectants are chiefly divided into:

Physical agents

Radiant energy like sunlight, ultraviolet light, cathode rays, electric current, etc. Heat, osmotic pressure, desiccation, salt effects, etc.

Chemical agents

A. Inorganic

1. Acids and alkalies.
2. Oxidizing agents—e.g., hydrogen peroxide, potassium permanganate.
3. Halogen compounds—chlorine, hypochlorite, iodine, etc.
4. Heavy metals—e.g., copper, mercury, silver, etc., and their salts as well.

B. Organic

1. Alcohol, ether, formaldehyde.
2. Coal-tar derivatives—phenol, cresol, etc., and aniline dyes.
3. Synthetic products—hexylresorcinol, acriflavine, synthetic detergents.
4. Essential oils.

Physical Agents

Sunlight

Sunlight has a definite bactericidal action. In this respect, direct sunlight is better than diffuse daylight. The lethal power is due to the action of the short wave spectra in the region of the ultraviolet ray, between, 2,000 and 2,950 Å, with a maximum effect at 2,650 Å.

Ultraviolet light

The power of penetration of ultraviolet light is, however, low so that the action of these rays is not definitely known, but it has been recently shown to be due to production of peroxides.

It has been observed that the lethal action of these rays may be increased by addition of some fluorescent dyes like cosine, methylene blue etc., to the bacterial suspension. The rays that are relatively inert under ordinary conditions become bactericidal in presence of these agents and the action is known as *photodynamic activity*. Ultraviolet light from mercury vapour lamp has been used for sterilization of operation theatres or water in a laboratory.

Cathode rays

Cathode rays or X-rays are destructive to tissue cells, but they have no marked bactericidal effect. Cathode rays, however, have been found to kill *Eash. coli. Salm. typhosa* and *Staph. aureus*.

Electric current

Electric current has been tried as a means of sterilizing fluids like water, milk, etc., but the results are not satisfactory. The current carries the organisms to the positive pole, as the bacteria are negatively charged. Some bactericidal actions have however, been attained by electricity. This can be attributed to the production of heat, H_2O_2 or acid in the medium rather than any direct action of the current.

Radium

It can kill organisms or cause attenuation of virulence depending on the time and intensity of irradiation.

Heat

Heat coagulates the bacterial protein. It has been observed that moist heat is more effective than dry heat. Sterilization by steam is an example to moist heat, whereas, dry heat in a hot air oven is less effective and requires a higher temperature for destruction of bacteria than steam. The better action of moist heat is attributed to (a) its increased power of penetration, (b) abundant water, which can coagulate protein more easily, (c) steam, which is a better conductor of heat than air. Autoclaving, tyndallization, etc., have already been dealt with in the chapter of sterilization.

Osmotic pressure

Concentrated solutions of salts and sugars are used to preserve substances against bacterial growth.

Desiccation

Drying has a deleterious effect on bacteria and vegetative organisms can hardly withstand drying; but spores often resist this condition. On the contrary, cold and desiccation preserve bacteria and type cultures in the laboratory are specially preserved nowadays by freeze-drying *in vacuo*, which is known as *lyophile method*. When preserved in such a condition, some of the organisms remain viable for a limited time and retain their virulence without any alteration of the antigenic structure.

Sonic and supersonic waves

These have been utilized to kill bacteria. Sonic waves have audible frequency, whereas supersonic waves are above audible frequency. Violent agitation, produced by these waves, causes disintegration of the bacterial cell with liberation of their antigenic constituents.

Inorganic Chemical Agents

Distilled water

Neutral distilled water of pH 7 has been found to be detrimental to bacteria, but the action varies differently with different bacteria under different conditions. The action is marked, when the H ion concentration is increased to pH 4; the actio is also marked, when the pH is 8 or above. Spores, however, survive for 4 to 5 in natural water of streams and lakes, which are really complex solutions containing some nutrient.

Salt effect

In low concentration, inorganic salts help the growth of bacteria, but as the concentration is increased, the salt becomes toxic and injures the cell. Bacteria are often resistant to osmotic changes produced by these salts, but the inhibitory effect is chiefly attributed to.

(a) Change of pH due to the action of electrolytes upon cell membrane,

(b) Influence on the permeability and physical state of the protoplasm,

(c) Interference with indigenous metabolism,

(d) Interference with collular enzymic activity.

Acids

Inorganic acids are more effective bactericidal agents than organic ones. The action is largely dependent upon the electrolytic dissociation of H ions. In this respect, strong acid, like hydrochloric acid is more effective than weak acid as acetic acid.

Alkalies

The action of alkalies, as disinfectants depends on the electrolytic dissociation and concentration of OH ions. But, in this respect, H ions are relatively more effective than OH ions. The disinfecting action of soap is mainly due to the presence of alkalies.

Oxidizing agents

Chemicals, which readily give up oxygen in presence of organic substances, are effective disinfectants. The most useful oxidizing agents, commonly used are hydrogen peroxide and potassium permanganate. Hydrogen peroxide liberates oxygen when it comes in contact with pus and this nascent oxygen acts as a mild antiseptic. Weak oxidizing agents convert 'S-H' group to oxidized 'S-S' form, while stronger ones convert it to irreversible stage of oxidation.

Halogens

Chemicals of this group, namely, chlorine, bromine and iodine are effective disinfectants. Their actions and toxicities are inversely proportional to their atomic weights. The action is partly lost owing to the fact that, they break down rapidly in aqueous solutions. They cannot be used in high concentration also, as they damage the organic matters. In practice, chlorine and iodine are commonly used as antiseptics. Chlorine is extensively used for disinfection of water, but as it combines readily with organic matters, its action is deflected by the presence of such materials in water. As disinfectants, chlorinated lime and chloramines are used. Eusol and Dakin's solution which are similar in action are used in sloughing ulcers with putrefying condition. Eusol is derived from simple neutralization of calcium hypochlorite with boric acid and is a useful non-irritant antiseptic. Tincture iodine (Liquor iodi mitis) is effective and non-irritant as a disinfectant of the skin and has stood the test of time.

Heavy metals and their salts

Metals with lower atomic weights are less toxic to bacteria than those with higher atomic weights. The action is inhibitory in lower concentration, but in higher concentration they are destructive.

Oligodynamic action of metals

Extremely diluted solutions of heavy metals inhibit the growth of bacteria. Copper, mercury and silver have such effects.

Copper

Copper in a dilution of 1 in 10 distilled water kills *Salm. typhosa* in 2 hours, but the action is less marked in tap water or sea water.

Mercury

The action of mercuric chloride on bacteria is complex and occurs in two stages. The metal is first absorbed on the surface of bacteria and then it enters and kills the organism. The action is due to mercury ions of mercuric chloride and it acts by combining with the 'SH' group of the cell. It kills Salm. typhosa in a dilution of 106, in 24 hours, but in a dilution of 103, it kills the organism in two and half minutes. Presence of protein, like blood or serum in the medium markedly inhibits its action. According of Chick and Martin (1908), presence of 3 per cent faeces in a solution of $HgCl_2$ reduces the disinfecting action by 90 per cent.

Organic salts of mercury like phenylmercuric nitrate, mercurochrome (di-sodium salt of 2, 7-dibromohydroxy-mercurifluorescein) and merthiolate (sodium ethylmercuri-thiosalicylate, containing 49 per cent of mercury) are used in practice. Mercurochrome is used on the mucous surface as 2 per cent solution, but merthiolate is used in 1 : 5,000 to 1 : 30,000 dilutions. Phenyl mercuric nitrate is a good antiseptic and used as a preservation of sera and vaccines in a dilution of 1 : 50,000. Phenyl mercuric nitrate or acetate is a powerful fungicide and used in ringworm infection.

Silver

It is also an effective antiseptic in a high dilution. Silver nitrate, 1 : 10,000 inhibits bacteria. In one per cent solution, it is instilled routinely into the eyes of the new-born infants as prophylactic against *ophthalmia neonatorum*. The action is markedly less in presence of protein, when the kation acts on the proteins to form an insoluble albuminate. The salts usually used are silver proteinates like argyrol and protargol. The *oligodynamic effect* is probably due to free silver ions released in low concentration from the metal. This bactericidal effect of silver is made use of in the Catadyne process of water disinfection used for domestic installation. The action is less marked in presence of protein than in distilled water. Gram-positive organisms are more susceptible to the action of metals or their salts than Gram-negative ones.

Organic Chemical Agents

Alcohol

Absolute ethyl alcohol is often used as an antiseptic, but 50 to 80 per cent a alcohol has a better disinfectant action than absolute alcohol. Germicidal action of alcohol increases with the increase of its molecular weight. Thus the action increases with increase of the molecular weight, e.g., methyl, ethyl, propyl and butyl alcohol in that order. Ether-alcohol mixture is usually applies for sterilization of skin. Ether cleans the surface by removal of fat, then alcohol acts as a disinfectant.

Formaldehyde

It is a potent antiseptic and the action of commercial formalin is due to the presence of formaldehyde (40%). It fixes the body tissues and it is better used as a fixative. But it is also used o disinfect rooms and for this purpose paraform tablets are used; 20 gms are required to disinfect a room of 1000 cu ft.

Coal-tar derivatives

(Coal-tar products like phenol, cresol, etc., are powerful disinfectants. They are insoluble in water except phenol. In colloidal suspensions, they act by adsorption on the bacterial cell, followed by penetration into the cell.

Phenol

It is bacteriostatic in a dilution of 1 in 1000, but it is bactericidal in 1 to 2 per cent aqueous solutions and kills almost all-vegetative bacteria. The action is due to coagulation of protein. Phenol in 0.5 per cent strength is used as preservative of vaccines and sera.

Cresol

This is the simplest of alkyl phenol. Various isomers of cresol like *ortho-*, *meta* or *para*-cresols are usually used as a mixture in tricresol, It is four times stronger than phenol. Tricresol emulsified with green soap (3 per cent), forms lysol. A solution of 2 per cent lysol is used as a disinfectant in the laboratory to receive contaminated plugs, pipettes, etc. Tricresol in 0.15 per cent strength is used as a preservative of vaccines and sera.

The antibacterial activity of phenol is increased by various substitutions I the phenol nucleus, namely, *alkyl* and *chloro*—derivatives or, *diphenyls*; many of them having high phenol coefficients of 100 or more. *Simpler compounds* of phenol are active against Gram positive and Gram negative organisms and are very little affected by organic matters, whereas high molecular weight compounds are more active against Gram positive than Gram-negative ones, but they are less effective in presence of organic matters.

The xylenols or methyl cresol include two important disinfectants, namely, *p-chloro-sym-m-xylenol* (PCMX) and 2 : 4 dichloro-*sym-m-*

xylenol (DCMX), the former was frequently used in commercial disinfectants in the past. PCMX has a marked antibacterial effect against both Gram-positive and Gram-negative bacteria.

The important diphenyls are the chloro-substituted diphenylmethanes, e.g., dichlorophane, tetrachlorophane and hexachlorophane.

Chlorhexidine (1: 6—di-4'-chlorophenyl-deguanidohexane) has been extensively used in recent years as skin disinfectant. It is of low toxicity and active against Gram-positive and Gram-negative bacteria. A solution of 0.5 per cent digluconate in 70 per cent alcohol is as effective a skin disinfectant as tincture iodine, but without any irritating effect. It is also used in a cream base for the treatment of nasal carriers of *Staph. aureus*.

Aniline dyes

Basic aniline or coal-tar dyes, namely, triphenylmethane dyes in the form of crystal-violet, gentian-violet, etc., have marked action an Gram-positive organisms, but they have little action on gram-negative bacteria. Tubercle bacillus is not inhibited by 1 in 10,000 dilution of gentian-violet. Dye like brilliant green is as effective as crystal violet. Brilliant green is used as an inhibitory substances in selective media for isolation of *Salm. typhosa*. It inhibits *Esch. coli* in a concentration of 1 : 500,000.

Synthetic Products

Hexylresorcinol

It is a strong disinfectant and kills vegetative organisms in 15 seconds; it also kills spores after a longer contact. It is available as 1 in 1000 solutions in 30 per cent glycerine in water.

Acridine dyes

Proflavine and acriflavine usually inhibit Gram-positive and to a lesser extent Gram-negative bacteria. They are non-toxic to the tissue. Since they combine with nucleotide, it is likely that they may interfere with coenzyme system of bacteria and block certain metabolic paths. They act best at a certain *p*H near about 7.

Synthetic detergents

The main action of these substances is due to the presence of long-chain fatty acids, which have the property of concentrating at the bacterium-water interfaces. They are, therefore, called *surface active agent* or detergents. Detergents concentrate at the delicate cell membrane and thus they act by disrupting the normal function of the cell membrane. Some detergents denature protein, as they 'leak' soluble nitrogen and phosphorus compounds in the medium. They may also inhibit the action of enzymes. Detergents are good wetting and cleaning agents and destroy bacteria. They are relatively non-irritant to raw surfaces, but may be toxic to leukocytes and may denature tissue protein. Their penetrating power is low and their activity is greatly reduced by the presence of organic matters.

Essential oils

Essential oils like cinnamon oil, clove oil, sandalwood oil, etc., act as disinfectants. They are bactericidal in high concentration but in a lower concentration, they are bacteriostatic.

Factor which modify the action of disinfectants :

(a) Concentration of the disinfectant. Stronger solutions are more effective than weaker ones.

(b) Medium in which it acts. Presence of organic matters interferes with disinfecting action to a great extent.

(c) Time of exposure. Longer the time of exposure, better is the germicidal action.

(d) Temperature. A disinfectant is more effective at a higher temperature than at a lower temperature.

(e) Type of bacteria used. Disinfecting action is more effective o vegetative organisms, but less effective on spores and spore-bearing organisms.

Dynamics of disinfection

The phenomenon of bacterial death by disinfectants is a complex interrelated, biological process and cannot always be determined by mathematical means. Quantitative studies o the rate oat which the organisms die have been represented graphically. If the logarithm of the number of viable organisms after the action of a disinfectant are plotted against time, the points tend to fall in a straight line. This is applicable both to the death of spores and the vegetative cells under the influence of physical agents or chemical disinfectants.

The velocity of the reaction depends on the kind and concentration of the disinfectant, nature of organisms, whether vegetative or spores and other factors, when influence the process of disinfection. The reaction is expressed mathematically in the following formula : $K=1/t \log B/b$, where B=initial number of organisms, and b=the number after t time. From the logarithmic curve the process is considered to be mono-molecular chemical reaction and therefore, follows the law of mass action, where the reacting substances are bacteria and disinfectants. Experimental evidences indicate that, such logarithmic plots may be compared with a straight line in some cases only, when relatively high concentrations are used, but in other cases, a sigmoid curve is obtained, the death rate being more rapid in the beginning is some but. Less so in others, or even irregular. The velocity of reaction of disinfectants, however, depends on the temperature and pH at the time of reaction. An increase of 10°C increases the reaction velocity of phenol 8 times and mercuric chloride 2 to 4 times. The acidity of the medium also increases the reaction velocity.

Standardization of disinfectants

Disinfectants may be standardized *in vitro*; it is known as phenol coefficient (Rideal-Walker) test. This is quite different form its action in vivo and does not represent any accurate indication of their activity in presence of animal tissue. For this purpose, Chick-Martin test is carried out in presence of organic matters. Besides this, there are several other coefficients for the determination of antiseptic properties.

Inhibition coefficient (Bacteriostatic property)

It is the lowest percentage concentration of an antiseptic that completely inhibits the growth of the organisms in nutrient agar. Bacterial suspension is prepared in distilled water and nutrient agar containing phenol is inoculated with this suspension. There will be no growth of *Salm. typhosa*, usually in a concentration of 0.25 per cent, which indicate the inhibitory concentration of phenol. Other antiseptics may similarly be tested and the coefficient is drawn in relation to phenol.

Method

The antiseptic to be tested is incorporated in different concentrations in 10 ml amounts of nutrient agar. The medium is poured into plates. Stroke cultures are then made from the bacterial suspension in sterile distilled water showing a faint turbidity to the naked eye. The inoculated plate is incubated for 48 hours after which reading is taken for the inhibition of the growth. Several organisms may be tried at the same time by making stroke inoculations on each of the plates.

Instead of solid medium, broth with carefully adjusted pH or serum (heated to 56°C for several hours) may be used for the test. In such a case, varying concentrations of antiseptics are used in the medium and each tube is inoculated with a standard amount of bacterial suspension.

The tubes are incubated for 48 hours at 37°C. Subcultures are then made on agar tubes, one tube for each concentration. The lowest concentration, in which growth will be absent, will indicate inhibitory concentration and the index of the antiseptic is obtained by comparing the result with hat of phenol.

Inferior lethal coefficient (Bactericidal property)

It indicates the concentration of an antiseptic and time of exposure required to kill non-sporing bacteria. This is done by preparing a series of dilutions of the antiseptic in a fixed quantity of sterile distilled water and adding a constant quantity of bacterial suspension. A loopful of the material is then transferred at intervals to nutrient broth or in melted agar at 45°C for preparation of poured plates. The quantity of antiseptic, transferred over to the culture medium, must be less than the inhibition coefficient. The broth cultures of the poured plates are then incubated for 48 hours. Absence of growth after 48 hours indicates lethal concentration of the antiseptic in given time.

Superior lethal coefficient

This expresses the concentration of a disinfectant and the time of exposure required to kill bacterial spores. A solution of 4 per cent potassium permanganate kills the spore of *B. anthracis* in 20 minutes.

Phenol coefficient

It expresses the disinfecting power of a particular substance as compared to that of pure phenol. In other words it may be expressed as:

1. Dilution of disinfectant which sterilizes the suspension in a given time.
2. Dilution of phenol which sterilizes the suspension in the same time.

Requirements

1. Lab-Lemco broth-Lab.Lemo-20gm, eupeptone 20 gm, sodium chloride-10 gm and distilled water–1,000 ml; pH 7.6. Distribute exactly in 5 ml amounts into test tubes and sterilize in the autoclave.
2. Standard loop, made of 28 S.W.G. platinum wire with an internal diameter of 4 mm, bent almost at right angle to the wire, so that eye of the loop remains horizontal.
3. 5 per cent solution of pure phenol crystals (crystalling at 40.5°C). This is standardized with decinormal bromine. It keeps indefinitely

in stoppreed bottle. Dilutions of various working strength are made by diluting it with sterile distilled water.

4. Broth culture prepared from a standard culture of *Salm. typhosa*. Subcultures are made in standardized broth for 3 subsequent days and then 24 hours old broth culture is used for the test.
5. Special test tube rack. (a) It consists of two tiers: the upper tier contains 20 holes in 2 rows to hold test tubes 3" by ¾"; each row containing two sets of five. They are marked 1 to 20. (b) The lower tier contains the dilution tubes. This is a mental rack consisting of 5 holes for holding long tubes which are dipped into a copper water bath at 18°C all through the test. These will contain the dilutions of the disinfectant from higher to lower dilutions.
6. Sterile dropping pipettes for measuring 0.5 ml of broth culture.
7. Sterile dropping pipettes for measuring 0.5 ml of broth culture.
8. Sterile graduated pipettes and stoppered measuring cylinders for preparing suitable dilutions of the disinfectant to be tested.
9. Sterile test tube 3" by ¾" and larger sized test tubes for using as dilution tubes.

Method of testing

1. The inhibition coefficient of the disinfectant is previously determined for the standard strain of *Sam. yphosa*. Dilute the disinfectant in five graded-dilutions, the highest being stronger than the inhibition coefficient.
2. From the 5 per cent (1 to 20) stock solution of phenol, dilutions of 1 in 95, 1 in 100, 1 in 105, 1 in 110 and 1 in 115 are freshly prepared.
3. A volume of 5 ml of each dilution of the disinfectant to be tested is distributed in each of the five dilution tubes. The dilutions are prepared accurately with distilled water.
4. Add with a sterile pipette 0.2 ml of 20 hours' broth culture of Salm, typhosa to the tubes of the dilution rack at interval of half a minute and mix each time after adding the suspension.
5. The dilution tubes in the rack are kept (during the test) in a water bath at 18°C. When the culture has been added to the fifth dilution tube, the first tube has been exposed to the disinfectant for 2 minutes. Take the sterile loop and in another half a minute, a loopful of the material is taken from the first dilution tube and inoculated in broth tube 1. A loopful of the material is brought

out half a minute later from the dilution tube 2 and inoculated into broth tube 2 and so on, till inoculations are complete from all the five dilution tubes. Next come back to the dilution number 1 and inoculate into the tube 6 after half a minute and so on. In this way, the process is continued at intervals of half a minute till all the twenty broth tubes have been inoculated. The material from 5 dilutions of the disinfectant are thus inoculated at intervals of 2½, 5, 7 and 10 minutes.

In the same way, different dilutions of phenol, e.g., 1 in 115, 1 in 110, 1 in 105, 1 in 100 and 1 in 95 are, tried in a separate unit. The inoculated broth tubes are then incubated at 37°C for 48 hours and examined for growth as shown by turbidity.

Results are recorded according to the growth in the broth tubes. Phenol coefficient is obtained by dividing the highest dilution of the antiseptic in which growth was inhibited with that of phenol in a given time. From the above test, the phenol coefficient is 900/100 in 7½ minutes=9.

Treating Diseases with Antimicrobial Agents

The control of disease-causing microorganisms is fundamental to the practice of modern medicine. Today there are a large number of anti-microbial agents available for treating diseases caused by microorganism. Such drugs have enabled physicians to treat many disease conditions that a few years ago were frequently fatal. Many infectious diseases are relatively easily treated with antibiotics. We need only think about the number of times a physician has prescribed an antibiotic for us or our children to appreciate the wide us of antibiotics in medical practice. Many once widespread deadly diseases, such as plague, are rare today because of the effective use of antibiotics. The use of antimicrobial agents for treating infectious diseases has led to greatly increased life expectancies.

The antimicrobial agents used in medical practice are aimed at eliminating infecting microorganisms or at preventing the establishment of an infection. These chemicals should not be confused with the large number of drugs used in medical practice for alleviating the symptoms of disease or for treating diseases not caused by microorganisms. To be of therapeutic use, an antimicrobial agent must exhibit selective toxicity. To have therapeutic value, and antimicrobial agent must inhibit infecting pathogens than to the host organism. A drug that kills the patient is of no use in, treating infectious diseases, whether or not it also kills the pathogens ! As a rule, antimicrobial agents are of most use in medicine when the mode of action of the antimicrobial chemicals involves biochemical features of the invading pathogens not possessed by normal host cells.

Antibiotics

Antibiotics are biochemicals produced by micro-organisms that inhibit the growth of, or kill, other microorganisms. The discovery and use of antibiotics have revolutionized medical practice in the twentieth century. By their very nature, antibiotics must exhibit selective toxicity because they are produced by one microorganism and exert varying degrees of toxicity against others. The formal definition of an antibiotic distinguishes biochemicals that are produced by microorganisms from organic chemicals that are synthesized in the laboratory. This distinction is no longer meaningful because organic chemists can synthesize the biochemical structures of many naturally occurring antibiotics.

Additionally, many antibiotics in current medical use are chemically modified forms of microbial biosynthetic products. Consequently the term antimicrobic is now widely used in medical fields instead of the term antibiotic. In this chapter both the terms antibiotic and antimicrobic are used to refer to both natural and semisynthetic antimicrobial agents that are available to the physician for treating infectious diseases.

Antibiotics Effective Agents Bacterial Infections

The biochemical differences in the cell structures of bacterial cells and eukaryotic cells form the basis for the effective use of antibiotics against bacterial infections. The bacterial cells with its unique murein-containing wall and 70S ribosome has two obvious sites against which antibacterial agents may be directed. Most of the common antibiotics used in medicine for treating bacterial infections are inhibitors of cell wall or protein synthesis. Some antibiotics are more selective than others with respect to the bacterial species that they can inhibit. A narrow spectrum antibiotic may be targeted at a particular pathogen, for instance at Gram positive cocci, or at a particular bacterial species. In contrast, some antibiotics have a broad spectrum of action, inhibiting a relatively wide range of bacterial species, including both Gram positive and Gram negative types. The choice of a particular antibiotic depends in part on the biochemical properties of the specific infecting bacterial strain.

In most cases, physicians make an educated guess as to which antibiotic is appropriate for treating a particular infection, and the selection of the antibiotic is based on the most likely pathogen causing the given disease symptomology and the antibiotics generally known to be effective against such pathogens. Many times a physician will select a broad spectrum antibiotic in order to ensure effective treatment. Only in special cases, such as when a patient fails to respond to a particular antibiotic and an infection persists, is an attempt usually made to isolate the pathogenic bacterium and to determine the range of specific antibiotic sensitivity of that organism.

However, concern is mounting in the medical field about the overuse of antibiotics because an undesired side effect of such "drug abuse" is the selection for disease-causing antibiotic resistant strains. It is now considered proper medical practice to perform culture and sensitivity studies to determine the proper antibiotic for treating a patient. Only in cases of life threatening infections should antibiotics be used without such testing to avoid the selective pressure for the development of antibiotic resistant pathogens. The importance of the problem of excessive use of antibiotics was recently underscored when the American Medical Association called upon physicians to avoid unnecessary use of antibiotics.

The reason for concern about how we use antibiotics is because numerous bacterial strains have acquired the ability to resist the effects of some antibiotics, with some bacterial stains, generally those containing R plasmids, having multiple antibiotic resistance. The basis of resistance in some cases rests with the ability of the particular strain to produce enzymes that degrade the antibiotics, preventing the active form of the antibiotic from reaching the bacterial cells where they could be inhibitory.

For example, some bacterial strains produce penicillinase enzymes (β-lactamases) that are able to degrade the antibiotic penicillin, making such strains resistant to penicillin. Resistance may also be due to decreased drug uptake, metabolic trans-formation of the drug to its nontoxic forms, decreased transformation of the drug to its active form, and/or decreased sensitivity of the microbial structure against which the drug is directed.

There is also concern about the cost of antibiotics prescribed by physicians. In many cases the selection of a alternate antibiotic in cases where the pathogen is resistant to the antibiotic of first choice greatly increases the cost of treating that disease. Additionally many antibiotics are produced by several companies under different trade name (Table 11.1). Many municipalities now require pharmacies to fill prescriptions by their by their generic names to encourage competitive pricing, unless a specific brand is specified. Such laws are controversial because the pharmaceutical manufactures claim they will lose the incentive to develop new antibiotics and to maintain high and costly quality control when a less stringently regulated and less expensive product will be used to fill a prescription. The public wants cost effective but quality antibiotics.

Table 11.1. Generic and trade name for some common antibiotics

Generic Name	*Trade Names*
ampicillin	Amcill, Omnipin, Penbriti Polycillin
cephalexin	Keflex
cephalothin	Keflin
chloramphenicol	Chloromycetin, Mychel
chloroquine	Aralen, Avloclor, Resochit
chlorotetracycline	Aureomycin
demeclocycline	Declomycin
doxycycline	Vibramycin
erythromycin	Ilotycin
gentamicin	Garamycin
kanamycin	Kantrex
methacycline	Rondomycin
minocycline	Minocin, Vectrin
nystatin	Mycostatin, Nilstat
oxytetracycline	Teramycin
penicillin G	Crysticillin, Duracillin
tetracyline	Aeromycin, Panmycin, Tetracyn, Tetrachel, Rexamycin
trimethoprim sulfamethoxazole	Bactrin, Septra

ACTING ON THE CELL WALL

Penicillin

Fleming observed that growth of *Staph. aureus* was suppressed round a contaminating colony of *p. notatum*, on an agar plate. He (1929, 1931, 1932) showed that in liquid media the fungus produced a substance, called penicillin, which inhibited the growth of Staphylococci, Str. Haemolyticus and other Gram positive organisms in a concentration, which was non-toxic to leukocytes. Ten years later,

Florey and his associates isolated it in sufficient quantities for therapeutic trial.

Properties

Penicillin is comparatively stable in acid than in alkaline solution. The loss of potency is delayed at a temperature below 10°C. The action of the drug is diminished be metallic ions, specially copper, lead, zinc, etc., but it forms salts with potassium, sodium and calcium, which are soluble in water.

Range of activity of penicillin is principally bactericidal of susceptible organisms but it is bacteriostatic in a lower concentration. It has a marked bacteriostatic in a lower concentration. It has a marked bacteriostatic action on cocci, Gram-positive bacilli, spirochaetes and various Clostridia. It shows a wide variation of sensitivity towards staphylococci and *N. meningitidis*, whereas, streptococci, *N. gonorrhoeae*, *C. diphtheriae*, *B. anthracis* and Actinomyces are susceptible to it. About 10 per cent of staphylococci (Pencillinase producing) are naturally resistant to penicillin, probably by a process of natural selection. *Str. Faecalis* is resistant to penicillin. Combined with sulphonamides, it has synergistic action in the treatment of infections, caused by organisms sensitive to either of the drugs.

Preparations

Penicillin has been isolated in different absorption spectra and are called F(1), G(2), X(3) and K. Penicillin 'G' is the most active of all; penicillin 'K' is rapidly destroyed by the body fluids and is, therefore, of little therapeutic value. Penicillins are derivatives of 6-amino penicillin acid. The most important is benzyl penicillin or penicillin G (crystalline), phenoxymethyl penicillin or penicillin V, which is acid resistant and can be administered by the oral route. PAM or procain penicillin G in oil with 2 per cent aluminum monostearate is active for longer period than penicillin.

It is, therefore, used in the treatment of syphilis and has almost replaced arsenicals due to its specific action against *Tr. pallidum*. It also acts against psittacosis-trachoma viruses. Benethamine penicillin G maintains the therapeutic blood level for 3-4 days after a single dose intramuscularly. Ampicillin is the most recent addition, having considerable activity against a wide spectrum of Gram-negative organisms like Salmonella, Shigella and some strains of Proteus, against which other preparations of penicillin are ineffective. But it is less active than, benzyl penicillin against staphylococci and streptococci.

Mode of action

The exact mode of action of penicillin is not known. Penicillin exerts is effect best on the actively dividing bacteria (log phase) and it irreversibly combines with some constituents of the bacterial cells. Penicillin acts by preventing the assimilation of glutamic acid through the cell-wall and probably also other amino-acids needed by Gram-positive bacteria for synthesis of bacterial protein.

According to Park and Strominger (1949), in *Staph. aureus* exposed to low concentration to penicillin, a nucleotide containing muramic acid and certain other amino acids are accumulated in the cell-wall. It was subsequently found by the same workers that the concentration of muramic acid, alanine, lysine and glutamic acid were the same in the both cell uridine nucleotide and the cell-wall mucopeptide. This implied that nucleotide in the cell, probably a precursor of mucopeptide, was not transformed into mucopeptide by the action of penicillin. The was directly confirmed, when it was found that penicillin prevented the incorporation of labelled amino acid into mucopeptide without interfering with the incorporation of cell protein.

It has also been observed that penicillin interferes with the synthesis of teichoic acid of the cell wall. As penicillin is fixed to the cell lipoprotein, it is quite probable that the primary site of action may be the cell membrane. Bacitracin and cycloserine also act in a similar way.

Resistant forms

Penicillin resistant forms appear in various ways:

1. *Culture in media containing penicillin.* When a strain of staphylococcus is cultivated in media containing gradually increasing concentrations of penicillin, the organism can no longer assimilate glutamic acid. It might altogether change the nature of its metabolism and as such, the organism could only synthesize an internal supply of glutamic acid.
2. *Production of Penicillinase.* Organisms like *Esch. Coli, B. subtilis* may resist the action of penicillin by production of Penicillinase and making penicillin inactive. Penicillinase is an enzyme that splits the lactum bond of penicillin, and makes it inactive. Since many staphylococci produce penicillinase, drug sensitivity test is important for this. Staphylococci are cured of penicillinase production by acridine, the trait is believed to be controlled by a cytoplasmic particle (plasmid).

3. *Appearance of variants*, i.e., appearance of mutants occurring spontaneously from a sensitive bacterial population. Thus a small number of the organisms, usually 1 to 2 per million of the originally sensitive strain, may breed selectively, survive and continue no multiply to form penicillin resistant variants.

Penicillin, unlike sulphonamides is not markedly inhibited by blood, pus or tissue fluids. In this respect, cysteine hydrochloride is highly antagonistic to penicillin but is not a satisfactory agent for inactivating penicillin in culture media. *International Unit*. One unit of penicillin is equivalent to the surface activity of penicillin, containing 0.6 microgram of pure crystalline sodium penicillin G (1 mg contains 1650 Oxford units). One Oxford unit per ml gives an inhibition zone of 24 mm, when standard Oxford strain of *Staph. aureus* is used for the test on an agar plate (Cup and Plate method). *Penicillin assay*. It is usually done by "cup and plate method". Nowadays disc method is used.

Penicillinase resistant penicillin, etc.

Staphylococci have become enormously resistant to penicillin mainly by development of penicillinase but penicillins with different side chains have been devised to be resistant to penicillinase. These are *methicillin* and *cloxacillin* which have low affinity for staphylococcal penicillinase and, according to Novick (1962), half life time of *methicillin in presence of staphylococcal penicillinase* is 360,000 times longer than that of an equivalent concentration of benzylpenicillin. *Cloxapenicillin is hydrolysed by staphylococcal penicillinase* at about twice the rate of methicillin. On the other hand, cloxacillin has an activity 10 times more than that of methicillin against penicillin sensitive strains. *Cephalosporin* C, isolated from Cephalosporium with a nucleus of 7-amino Cephalosporanic acid, is active against penicillinase producing staphylococci.

Ampicillin

It is the direct outcome of adding different side chains to penicillin. It is poorly soluble in water. It is acid resistant and can be administrated by mouth. It is 4-8 times more active than benzyl penicillin and acts against many species of Gram-negative bacilli but only slightly less active against Gram-positive bacteria are Neisseriae. Most strains of *H. influenzae* are sensitive to 0.25 μg/ml, *Salm tyhosa* from 1-2 μg, *Esch. Coli* from 2-8 μg/ml, proteus species and some strains of *Pr. mirabilis* from 4-8 μg/ml but Klebsiella and all other species are resistant except *K. pneumoniae*.

Adicillin

This is the first penicillin discovered, which has increased activity against Gram-negative bacilli. This is derived from a Cephalosporium-like fungus. It has particular action against *Salmonella* spp., and has a hope to act against typhoid carriers.

Bacitracin

It is derived from a strain of *B. subtilis* and is a mixture of then polypeptides. It has more or less the same spectrum as that of penicillin, but is considerably nephrotoxic. Bacitracin *in vitro* test inhibits streptococci of group A in 95 per cent cases and hence, used to differentiate them from other groups.

Acting on the Cell Membrane

Tyrothricin

It is a mixture of *tyrocidin* and *gramicidin*, the two polypeptides derived from Gram-positive, sporing bacillus, called *B. brevis*. It is bactericidal to Gram-positive and Gram-negative organisms. The chief difficulty in its use is that it is almost insoluble in water and is too toxic, except as a local application.

Tyrocidin is highly toxic for the cell-wall of bacteria, whereas gramicidin, which is active only against Gram-positive bacteria is bacteriostatic in action. It exerts is action by interfering with assimilation and metabolism of phosphates.

Polymyxin

It is a group of polypeptide antibiotics produced by *B. polymyxa*. Five polymyxins, e.g., A, B, C, D and E are known. Polymyxin B sulphate and Polymyxin E are basic polypeptides used in practice. Polymyxins are selectively active against Gram-negative organisms. They are particularly useful in the treatment of infections caused by *Ps. aeruginosa*. They are of little value against proteus group of organisms.

Polymyxins are actively bactericidal, as they form non-multiplying bacteria and show considerable surface activity, as a result of which, bacterial cytoplasmic membrane is damaged, resulting in leakage of essential cytoplasmic constituents. Polymyxin A, C, and D are nephrotoxic, but the purified polymyxins B and E can be administered by parenteral route. Colistin is either identical with or closely related to polymyxin E.

Blocking Protein Synthesis

Chloramphenicol (Chloromycetin)

It is derived from *Streptomyces venezuelae*, a soil bacteria, but after identification of pure crystals, it is being synthesized on a commercial scale.

Action

It is active when administered per mouth to give an effective concentration in blood, cerebrospinal fluid and tissue fluids. The drug is purely bacteriostatic and has no bactericidal effect even in a high concentration. Chloramphenicol and tetracyclines, unlike penicillin. Do not inhibit the synthesis of cell-wall mucopeptide. Studies with labelled amino acids show hat these compounds inhibit the incorporation of amino acids into protein by inhibiting the peptide bond formation but the precise mode of action of these two groups is likely to be different. RNA formed in presence of Chloramphenicol is heterogenous of low molecular weight and unstable.

Range of activity

Chloromycetin has a wide range of activity and acts both on Gram-positive and Gram-negative bacteria. It is the drug of choice in salmonella infections. It is also active against *Bord. pertussis, H. influenzae, Brucella* group of organisms, *Ps. Aeruginosa, Pr. Vulgaris*, various rickettsiae and some viruses. It is specially useful in enteric fever and influenza meningitis in children.

Doze

The doze is 25 to 50 mg per kg body weight. In practice, a daily doze of 2 to 4 gm is administered at an interval not later than 8 hours.

Undesirable side effects

It is less irritant to the gastrointestinal tract than tetracycline drugs. But the drug is toxic to the bone-marrow due to the presence of nitrobenzene ring, which affects the haemopoietic system. This is why the administration of the drug should be restricted only to salmonella infections in which other drugs fail to act.

Assay

Chloramphenicol can be assayed colorimetrically, based on the reduction of nitro-group into primary amine, followed by diazotization and coupling with N-(1-naphthy 1) ethylenediamine hydrochloride. By this method, the total quantity is estimated including both active and inactive degradation products.

It is stable and diffuses into agar media. For this reason, it gives a larger zone of inhibition in the disk test than that of terramycin.

Tetracyclines

These are broad spectrum antibiotics and are effective against a large number of both Gram-positive and Gram-negative bacteria, rickettsiae and PLG organism. Tetracycline is the parent compound and chlortetracycline (aureomycin and its derivative demethylchlor-tetracycline) and oxytetracycline (terramycin) are the substitutes.

Action

Like chloramphenicol the tetracyclines also interfere with protein synthesis. Tetracycline probably interferes with the transfer of amino acids from the amino acyl S-RNA to polypeptide. Experimentally the tetracyclines have been shown to inhibit the incorporation of leucine into protein of cell free systems from rat liver. For this action a greater drug concentration is necessary for *Esch. coli*. The chealating properties of the tetracycline molecules probably initiates the action of inhibition of protein synthesis. Whether this chealating activity influences the integrity of active ribosomes dependent upon magnesium for its proper function or neutralization by magnesium brings about a conversion of the free drug to an inactive complex affecting the enzymic activity is not definitely known.

Terramycin (Oxytetracycline)

Terramycin is derived from *Streptomyces rimosus*, a soil organism. It is an amphoteric substance and is stable in acid hydrochloride form. Its action is like that of aureomyc, and it is active like aureomycin by mouth. Although it is an exact counterpart of aureomycin, it is claimed that it is more effective than aureomycin or chloramphenicol in the treatment of Q fever.

Aureomycin (Chlortetracycline)

Aureomycin is derived from *Streptomyces aureofaciens*, which produces a golden yellow pigment. The drug is available as hydrochloride and is active against a wide range of bacteria.

Range of activity

It is active against *Staph. aureus, Str. Pyogenes. D. pneumoniae*, Haemophilus and Neisseria groups of organisms. Penicillin resistant strains of staphylococci are sensitive to aureomycin. *B. anthracis* is sensitive to it. *Esch. coli*, and Salmonellae are moderately susceptible to high doses, but Proteus group and *Ps. aeruginosa* are completely resistant. It is superior to streptomycin in brucellosis and melioidosis. It is efficacious in secondary syphilis, and in leptospirosis and relapsing fever it is superior to penicillin. Rickettsia group of organisms and virus of psittacosis and lymphogranuloma have been proved to be susceptible to it.

Erythromycin

The range of activity is similar to that of penicillin, but in addition, it is also active against *H. influenzae, Bord. pertussis, Br. melitensis* and Rickettsiae. But the efficacy in the treatment of infections against these organisms is still to be evaluated. It has bactericidal effect against these organisms, but it has marked capacity to produce resistant strains and it has an intermediate position between streptomycin and penicillin. Therefore, it should be reserved for the treatment, particularly of staphylococcal infection resistant to other antibiotics.

ACTING VARIOUSLY

Streptomycin

Streptomycin was discovered by Schatz and Waksman (1943) from the mould *Streptomyces griseus*, a soil streptomyces (actinomyces). Di-hydrostreptomycin is produced by the hydrogenation of streptomycin. Di-hydrostreptomycin is less toxic than streptomycin. Streptomycin hydrochloride is freely soluble in water.

Range of activity

Streptomycin is active in vitro against most Gram-positive cocci, many Gram-negative bacilli and *Myco. Tuberculosis*. It has no action of viruses. It is specially used for the treatment of tuberculosis, either pulmonary or miliary form of the disease. For this a prolonged course is necessary for 3 to 4 months in daily doses of 1 to 2 gm. It is also effective against *K. pneumoniae*, *Pr. vulgaris*, *Esch coli*, *H. influenzae, Bord. pertussis, Shigella, P. pestis* and *Br. tularensis*. It has been proved to be of little value against salmonella organisms.

Acquired resistance

The greatest draw-back with streptomycin therapy is that bacteria soon acquire resistance to its action. On the other hand, the therapeutic and the toxic doses are very near each other and resistance of the organism against the drug will place it out of therapeutic effectiveness. *Myco. tuberculosis* soon becomes resistant to the drug but simultaneous administration of *Para*-aminosalicyclic acid (PAS) avoids the tissue.

Mode of action

The exact mode of action of streptomycin is not definitely known, but like other antibiotics it is bacteriostatic. There is evidence to indicate that it acts by interfering with the oxidation of carbohydrate and amino acids and disturbs the oxalacetatepyruvate reaction of the sensitive organism. Its action is inhibited by cysteine and other compounds, containing SH group. The tolerance of the organism is increased in presence of blood, serum, pus, tissue, autolysates or inflammatory exudates, to an extent of 4 to 8 times.

More recently different types of action have been postulated for streptomycin. However, it requires a small amount of growth for demonstration of lethal effects. Controversy exists about the major site of lethal effects. These are (a) damage to the cell membrane, (b) control of septation, (c) cessation of protein synthesis, and (d) breakdown of RNA. Sport and Stainers (1961) visualized the bacterial ribosomes as the site of action of streptomycin and assumed that association of antibiotic to ribosomes of sensitive organism prevent attachment of messenger RNA to these particles and thereby making protein synthesis impossible.

Toxicity

Toxic manifestations are local transient pain, nausea, pyrexia and skin rashes, like papular or urticarial eruptions. Allergy may develop with prolonged treatment or contact with the drug in the case of medical personnel handling the drug. Allergic reactions are commoner with Streptomycin, which may be controlled by changing to dihydro-streptomycin. With prolonged use of streptomycin there is a greater tendency to vestibular damage, while with dihydrostreptomycin the tendency to cause deafness is slightly greater.

Activity

It is bactericidal for Gram-negative bacilli, including some strains of proteus and also for Gram-positive cocci. It has some action against Myco. tuberculosis. Its action is marked against proteus group of organisms in daily doses of 2 to 3 gm intramuscularly. It is also useful as a topical application to the skin and surface wounds. It may cause some renal damage and deafness.

Kanamycin is closely related to neomycin with which it appears to show complete cross resistance. It has toxic action for the eighth nerve and the kidney, but it is less so than neomycin. Proteus organism is sensitive to 1-8 mg/ml of kanamycin or neomycin. Kanamycin is the drug of choice in urinary infection with proteus organisms, but if the organism is not eradicated within a few days, it is liable to be resistant.

Neomycin

Neomycin sulphate is a complex substance of unknown chemical nature, isolated from *Streptomyces fradiae*. It is stable but poorly absorbed from the intestinal tract. It may therefore, be used for surface application and by the oral route as disinfectant of the intestinal tract prior to operation. After intra-muscular injections, it is readily absorbed, distributed in the body and slowly excreted in the urine. It is chemically related to streptomycin with which it may show some cross resistance.

Sensitivity Test

Before administration of antibiotics, it is desirable to see the sensitivity of the isolate organism with different antibiotic disks as a routine procedure. These disks can be prepared in the laboratory or Bacto unidisk for antibiotics of different colours are available, suitable for 10 cm Petri dishes. Carpeting of the organism is prepared in plate containing suitable medium (blood or chocolate agar); the disk is then applied and incubated under suitable condition of growth (aerobic, anaerobic or with CO_2). The reading is taken next day. The inhibition zone round the disk indicates sensitivity of the organism to the particular antibiotic.

Spectrum of Activity

It is convenient to classify antibiotics into 4 main groups, according to their range of activity, namely, Groups A, B, C and D.

Group A. Penicillin, erythromycin, novobiocin, bacitracin, etc.

Group B. Tetracyclines and chloramphenicol.

Group C. Streptomycin, neomycin and kanamycin.

Group D. Polymyxin B and colistin.

Group A is strongly active against Gram-positive cocci like staphylococci, streptococci, *D. pneumoniae*; Gram-negative coccus *N. gonorrhoeae*; Gram-positive bacillin Clostridia *B. anthracis* and actinomyces.

Group B. Members of this group are strongly active against Salmonellae, and moderately active against almost all Gram-positive and Gram-negative cocci and bacilli (broad spectrum antibiotics).

Group C. The members of this group are strongly active against *Esch. coli*, Proteus group and *Myco tuberculosis* moderately active against staphylococci and some salmonellae.

Group D. These antibiotics are strongly active against *Ps. aeruginosa*; moderately active against *N. gonorrhoeae*, salmonellae, *Esch. coli* and Proteus organisms.

Serum Killing Power

Another approach to determining the effectiveness of an antibiotic is to measure serum killing power (Schlicter test). Instead of adding dilutions of an antibiotic to suspensions of bacteria in a growth medium, a bacterial suspension is added to dilutions of the patient's serum. The ability of the bacteria to grow in the patient's blood is assessed by measuring changes in turbidity. Assuming that the patient is being treated with an antibiotic, no bacterial growth should occur. The break point in the dilutions where bacterial growth occurs reflects the concentration of the antibiotic in the patient's blood and the *in vivo* effectiveness of the antibiotic in controlling the infection. Inhibition at dilutions of the patient's serum or greater than or equal to 1:8 is considered an acceptable level.

12

Bacterial Diagnostics

Use of high copy number bacterial RNA offers several advantages in a diagnostics context compared with current deoxyribonucleic acid-based assays. The opportunity to only detect viable cells by targeting labile RNA transcripts may create an opportunity for "*real-time*" monitoring of pathogen load in response to a treatment regimen, while the natural amplification provided by the relative abundance of the RNA target compared with its corresponding gene opens a door to potential nonamplified direct detection technologies. In this chapter, a method is described to accurately quantify specific RNA transcripts and thus determine their potential utility as "*high-copy*" targets. The quantification method described also has application in gene-expression analysis.

Historical Overview

Polymerase chain reaction (PCR)-based diagnostic assays are well established in the food, environmental, and clinical microbiology sectors. The advent of real-time PCR has heralded an upturn in the speed and specificity of these assays. Paradoxically, the high sensitivity of this technology has potentially become an obstacle to its applicability in the clinical setting, especially where residual amounts of bacterial DNA persisting in a patient sample are present. The potential of RNA-based assays to circumvent this obstacle and only detect live cells has been described. High copy number bacterial RNAs have been investigated for their ability to identify bacterial species, and a useful first step in these analyzes is the determination of transcript copy number for potential RNA targets. Quantification of specific RNA copy number in a clinical, environmental, or food sample also may provide a means to rapidly monitor the number of viable bacteria present in the sample. The application of an RNA quantification methodology as part of a theranostic strategy to monitor bacterial load during an antibacterial treatment also may provide useful information that cannot currently be obtained by either measurement of deoxyribonucleic acid (DNA) levels in a sample or by more time consuming conventional microbiological approaches.

Quantification experiments typically are described as being one of three types, absolute, relative, and comparatitive. Despite optimistic claims to the contrary, all quantification experiments are "relative" only approaching the grail of being "absolute" as the quality of the standards improves. To provide an accurate reflection of in vivo copy number, an in vitro system must be devised to mirror the intracellular state or, as a compromise, to allow the conditions used for enumeration of 'unknowns" be replicated in the preparation of the standards.

Real-time PCR using 5′ exonuclease probe technology particularly is suited to the quantification of gene copy number. An intrinsic characteristic of real-time PCR technology is that the cycle at which the fluorescent signal crosses the detection threshold (Ct value) is proportional to the starting concentration of target material. Standard curves may be prepared using known amounts of synthetically generated target material. For estimation of RNA copy number in samples, an in vitro transcribed complementary RNA (cRNA) standard homologous

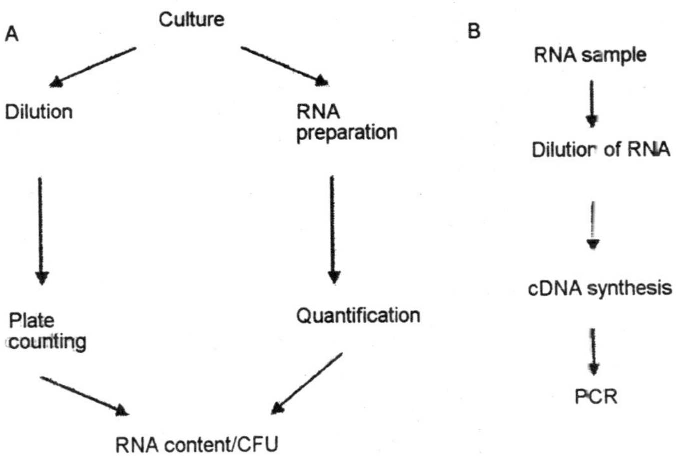

Fig. 12.1. Workflow of quantification experiment. (A) Calculation of recoverable RNA content per CFU. (B) Steps involved in quantitative reverse-transcriptase PCR from both standard curves and total RNA.

to the transcript being quantified is recommended for the generation of the standard curve.

Table 12.1. Comparison of DNA and RNA standard curves

Template	*CT value*
1×10^7 PCR products	22.25
1×10^7 RNA products	31.48
1×10^6 PCR products	25.73
1×10^6 RNA products	34.50
1×10^5 PCR product	29.10
1×10^5 RNA products	36.00

In a quantitative reverse-transcriptase (qRT)-PCR as with any quantitative analysis an essential element is the use of a "like-for-like" comparison between standard curves and the "unknown" material being analyzed. Because the materials being quantified in these experiments are RNA transcripts, the first step in this like-for-like analysis is the preparation of a set of artificially generated RNA transcripts for use in the standard curves. These transcripts are designed to be identical in sequence to the transcript being quantified. Therefore, a separate standard must be prepared for each RNA that is being quantified. A second level of complexity in the like-for-like analysis of standards and unknowns involves comparison of the amplification efficiencies of the various templates. Analysis of the mathematical models available for both RT-PCRs and PCRs demonstrate that even a slight fluctuation in reaction efficiencies can result in major differences in calculated number of initial RNA molecules.

To give an accurate representation of target copy number, the amplification efficiency of the standard curve should be as close as possible to the amplification efficiency of the unknowns. For example, if the amplification efficiency of the standard curve is high whereas there is some inhibition of amplification in the unknowns (resulting in a lower amplification efficiency), the quantification data returned will not be an accurate reflection of the target copy number. The standard curve depends in part upon the amplification efficiency of the standards. Therefore, every effort should be made to ensure that the amplification efficiencies of the standard curve and the unknowns are identical. Amplification efficiency (e) is calculated from the slope of a graph of the Ct values of a dilution series (either standards or unknowns) using the equation $e = 10^{-1/S}$, where S = slope.

In this chapter, a protocol is described for the quantification of high copy number RNAs in bacteria using in vitro transcribed cRNA standards and two- step qRT-PCR on the LightCycler instrument.

Materials

The quantification methodology described consists of four sections, in vitro transcription of cRNA for use in standard curves, preparation of unknowns (including determination of total RNA content per cell), two-step qRT-PCR protocols, and data analysis (including measurement of amplification efficiencies).

Preparation of cRNA for Standard Curves

1. Conventional thermocycler.
2. *Taq* polymerase and 10X PCR buffers.
3. dNTPs (Promega).
4. Heating blocks.
5. Invitrogen TOPO TA Cloning Kit.
6. 37°C incubators (Shaking and plate incubators).
7. Water bath.
8. Gel electrophoresis apparatus.
9. Ambion MAXIscript in vitro transcription kit.
10. Ambion NUCaway columns.
11. Agilent 2100 bioanalyzer and RNA 6000 Nano LabChip kit.
12. Ambion RNA 6000 ladder.

Preparation of Unknowns

1. Cell density meter.
2. RNA preparation.

Two-Step qRT-PCR

1. Conventional thermocycler.
2. 100 U/μL M-MLV reverse transcriptase.
3. 10X Reaction buffer: 500 m*M* Tris-HCl, pH 8.3, 500 m*M* KCl, 30 m*M* $MgCl_2$, 50 m*M* dithiothreitol.
4. dNTPs .
5. QIAquick PCR Product Purification Kit .
6. Real-time PCR machine (LightCycler was used for these experiments).

Data Analysis

We use the integrated LightCycler data analysis software.

METHODS

Preparation of cRNA for Standard Curves

The ultimate goal of the experimental design is to create a set of standards that mimic, as closely as possible, the conditions of the unknown sample being quantified. Therefore, in the case of an RNA copy number experiment, it is useful to use RNA in the preparation of the standard curve. At an early stage, the use of in vitro-transcribed RNA in the standard curve instead of quantified PCR products replicates the RT step carried out on the unknown samples.

Preparation of RNA standards involves the following steps:

1. Amplification by conventional PCR of target gene.
2. Ligation into suitable plasmid vector.
3. Transformation into competent cells.
4. Plasmid purification.
5. Screening to identify insert orientation relative to promoter site.
6. In vitro transcription.
7. Treatment with DNase.
8. RNA purification and quantification.
9. Dilution of RNA template to suitable concentration.

Amplification of the target gene

PCR amplification of the gene being quantified is performed using standard conditions. In the examples shown here a short (~200) bp fragment of the 16s ribosomal RNA (rRNA) gene from *L. onocytogenes* was amplified.

1. Prepare PCR mix for gene amplification in a 50-μL volume. The reaction mix should consist of 200 n*M* of each primer, 400 μ*M* dNTP, 5 μL 10X reaction buffer, and 1 U *Taq* polymerase.
2. Amplify the target DNA in a conventional thermocycler. In the case of 16S rRNA, amplification conditions are 1 cycle of 95°C for 5 min followed by 30 cycles of 95°C for 30 s, 50°C for 30 s, 72°C for 45 s, and a final extension step of 72°C for 5 min.

Ligation into plasmid vector

pCR II.1 TOPO cloning kit can be used for preparation of suitable clones. This plasmid has both T7 and Sp6 RNA polymerase promoter sites suitable for the in vitro transcription of specific RNA from the cloned insert. Follow the manufacturer's instructions for use.

Transformation into competent cells

1. Transform 1 μL of the ligation reaction into *E. coli* Inf' competent cells (they are included in the invitrogen kit) and incubate overnight on Luria-Bertani (LB) plates containing ampicillin and X-gal.
2. After an overnight incubation on LB agar plates containing X-gal, select eight white colonies and inoculate into 5 mL LB broths containing antibiotic.
3. Incubate overnight at 37°C with shaking.

Plasmid purification

1. Prepare plasmid DNA from the 5-mL overnight incubations using QIAprep Spin Miniprep Kit.
2. After purification, digest a 5-μL aliquot with *Eco*R1 at 37°C for 1 h.
3. Run the entire restriction reaction on 1.4% agarose gel.
4. Before proceeding to in vitro transcription, it is necessary to identify the orientation of the insert in the plasmid vector. To generate a sense RNA transcript, it is necessary for the 5′ end of the RNA to lie adjacent to the RNA polymerase promoter site. Sequencing of the cloned plasmid generates unambiguous data on the insert orientation as well as adding a confirmation that the correct gene was cloned.

In vitro transcription

To ensure that the cRNA produced by in vitro transcription is of a defined length, plasmids should be linearised using restriction enzyme *Hin*dIII.

1. Incubate plasmid in 50-μL volume at 37°C for 1 h.
2. Purify the plasmid using QIAquick PCR product purification kit and run a 10-μL aliquot on 1.5% agarose gel to confirm complete digestion.
3. Conducte in vitro transcription using 1 μg of plasmid DNA as template with the MAXIscript in vitro transcription kit for 1 h at 37°C.

Treatment with DNase

After incubation, DNase treat the samples using the DNase 1 *DNAfree*.

Purification and quantification of RNA

1. Remove free nucleotides, that is, dNTPs and rNTPs, by gel filtration over Nuc-Away Spin Columns.

2. Quantify 1-μL aliquots of the RNA on the Agilent 2100 bioanalyzer with RNA 6000 Nano LabChips.
3. Store RNA at –20°C until use in the RT-PCR.

Dilution of RNA template to a suitable concentration

Prepare 10-fold serial dilutions of DNase-treated RNA to span the range of concentrations from 1 ng/μL to 10 fg/μL.

Preparation of unknowns

Preparation of unknowns for quantification involves the following steps:

1. Determination of total RNA content/cell.
2. Total RNA preparation from appropriate growth phase.
3. Quantification of RNA.
4. Treatment with DNase.
5. Dilution of RNA to appropriate concentration (typically one cell equivalent).

Determination of total RNA content/cell

1. Prepare a standard curve of bacterial growth to identify stationary and exponential phases of growth.
2. Grow cultures to appropriate growth phase (typically mid-exponential phase and stationary phase after 24 h).
3. Prepare a 10-fold serial dilutions of culture by adding 20 to 180 μL of sterile broth. Dilutions should be prepared in triplicate.
4. Prepare appropriate dilutions and spread onto labeled agar plates. Incubate overnight at 3 7°C. Remove duplicate 5-mL aliquots from the culture and prepare total RNA using High-Pure viral RNA purification kit. After purification, quantify the RNA by measuring triplicate 5-μL aliquots of each preparation in a 500-μL volume at 260 nm in an ultra violet spectrophotometer.
5. Calculate the mean RNA content of the six measurements and extrapolate to the total RNA extracted from the 5 mL of culture.
6. After the overnight incubation, count the colonies on the plates.
7. Calculate the mean of the triplicate samples and extrapolate the CFU/mL of culture using the dilution factor. Calculate the mean RNA content/cell by dividing the total RNA per milliliter of culture by the total CFU/mL.

Preparation of total RNA

1. Incubate cultures for appropriate length of time to reach mid exponential and stationary phases of growth.

2. Remove duplicate 5-mL aliquots from cultures and pellet bacteria by centrifugation.
3. Discard the supernatant and resuspend cells in 200 jiL buffer before applying to Ribopure Yeast kit (recommended) or to High Pure Nucleic Acid purification columns.

Quantification of RNA

1. Duplicate 1-μL aliquots of total and in vitro-transcribed RNA should be quantified on the Agilent 2100 Bioannalyzer and RNA 6000 Nano LabChip system.
2. Alternatively, RNA may be quantified using optical density readings at 260 nm or by the use of RNA-specific dyes.

Treatment with DNase

Treat samples with DNase using the *DNAfree* protocol.

Dilution of RNA to appropriate concentrations

1. Prepare 10-fold serial dilutions by adding 2 μL of RNA to 18 μL of RNase-free water.
2. Serial dilutions from containing RNA equivalents of 10 to 1000 cells should be used to overcome potential PCR inhibitor effect of reverse transcriptase at low template concentration.

Two-Step qRT-PCR

RT

1. Incubate 1 μL of each dilution at 80°C for 3 min with 100 ng of reverse primer in 12 μL of volume.
2. Quench on ice before adding reaction mix containing 1 μL of M-MLV Reverse Transcriptase (100 U/μL), 2 μL of 10X reaction buffer, and 4 μL of 2 m*M* dNTP mix.
3. Conduct reverse transcription at 42°C for 1 h.
4. Denature RNA/DNA hybrids and RT enzyme at 95°C for 10 min.
5. Store cDNAs at –20°C.

Real-time PCR of cDNA

PCR amplification and real time detection was performed on the LightCycler instrument with TaqMan probes.

1. Prepare 20-μL reaction volumes for each sample using amplification conditions optimized for the gene transcript being quantified.
2. Templates should include standard curve cDNAs to span the range of expected target transcripts. Typical standard curves should contain dilutions containing from 1×10^3 to 1×10^7 copies.

3. Templates from total RNA should include cDNA prepared from a dilution series corresponding to from 10 to 1 $\times$ 10^3 cell equivalents.

Data Analysis

For quantification analysis, design standard curves to span the expected range of RNA copy numbers. Include in each experiment dilutions corresponding to RNA copy number from 1 $\times$ 10^3 to 1 $\times$ 10^7. Reverse transcribe unknown RNA corresponding to known cell equivalents and amplify by real-time PCR in parallel with the standards.

1. Integrated LightCycler data analysis software is used to plot a graph of the Ct values of both standard curve and dilutions of total RNA.
2. The slope of this line(s) can be used to calculate the amplification efficiency of the reactions (*e*) by application of the equation $e = 10^{-1/S}$.
3. Amplification efficiencies of standards and unknowns must be as close as posible for accurate quantification data to be returned.

Notes

1. RNA standards are prepared by in vitro transcription from a T7 promoter that has been attached to the 5´ end of the target gene. This generates a homogenous population of RNA molecules of known length. This material is quantified and the number molecules/ microliter calculated. Serial dilutions of this RNA provide the template for the generation of an RT-PCR standard curve.
2. Any RT-PCR experiment that uses dilutions of RNA as starting material also will be influenced by inhibition of the RT reaction. The alternative approach, preparing a large stock of cDNA and diluting down from, this will not display this problem. There may be inhibition in the single large cDNA reaction but, because the standard curve is not designed to measure RT efficiency, just the subsequent PCR, its effect remains unnoticed. Amplification efficiencies are calculated by measuring the slope of a plot of Ct values for a dilution series of template, $e = 10^{-1/S}$.
3. Short (~200 bp) PCR products from the genes of interest are cloned into plasmids containing a T7 RNA polymerase promoter site. In vitro transcription is used to generate RNA transcripts consisting of the cloned fragment and a short plasmid sequence. These transcripts are then column purified and quantified by ultraviolet spectroscopy. One 200-base RNA transcript is calculated to be equivalent to 0.2 ag (1 ag = 1 $\times$ 10^{-18} g) of this RNA.

Serial dilutions of the in vitro-transcribed RNA are used as template for RT, and this cDNA is amplified using the Light-Cycler.

4. An alternative approach to generate starting template for in vitro transcription involves using PCR where the forward primer has been modified to include the T7 promoter sequence (taatacgactcactatagg). We chose the cloning strategy described here because it generates a stable pool of material from which fresh template can be prepared.
5. T7 RNA polymerase is highly processive and will continue transcription until a stop signal is encountered. In the case of a circular plasmid, the polymerase could potentially continue indefinitely. The restriction digestion prevents this and also results in the RNA transcripts being of uniform known size, which eases quantification calculations.
6. Several methodologies suggest purification of the in vitro-transcribed RNA immediately after transcription. However, it is our experience that it is advisable to perform DNase treatment of the prepared RNA *before* column purification. This step not only reduces the overall number of purification steps by one (it is no longer necessary to conduct a separate post-DNase purification) but it also facilitates the column purification of both free nucleotides and ribonucleotides from the mix in one step. Column purification after DNase treatment also has been shown to improve the accuracy of spectrographic readings taken to quantify nucleic acid concentration.
7. Several online calculators are available to determine the molecular weight of an RNA transcript. A 200-base RNA will have an average molecular weight of 64 kDa, which corresponds to a single 200-base transcript having a mass of 0.1 ag or 1×10^{-19}g
8. Because of the large numbers of plates that will be involved, it is recommended that an initial "ranging" experiment be carried out to identify which dilutions will result in countable plates. Prepare 10-fold dilutions and plate 200 mL onto plate. Incubate overnight at 37°C. Examine the plates to identify which of the dilutions result in between 30 and 300 CFU/plate. These are the dilutions that should be used in the triplicate plate count experiment.
9. The figures generated for amount of RNA per cell are limited by several factors. First, the CFU per milliliter may underestimate the total number of cells present in the original culture as not all

cell produce a colony. Second, no RNA preparation method can be guaranteed to result in 100% purification of RNA from cultures. Therefore, we apply the term "recoverable RNA yield CFU^{-1}" to more accurately describe our starting material.

10. At low concentrations of template, RT inhibits subsequent PCRs. The inhibitorary effect of RT on PCR is removed at template concentrations beyond 10^5 copies. When designing standard curves, it is useful to avoid lower dilutions of template because the amplification efficiency calculations will be affected by this inhibitory effect. An alternative solution mentioned also may be useful.
11. We also have investigated the possibility of using a dilution series of cDNA samples to generate the standard curves. Provided that the amplification efficiencies of these samples match those prepared from the RNA dilution series, this method provides several advantages. First, the total number of RT reactions is reduced; second, the cDNA template used to prepare the standards is more stable than an in vitro-transcribed RNA; third, this method is more accurate at lower template dilutions because it eliminates an inhibitory effect of components of the RT reaction on the PCR amplification. We found that diluting the high copy number cDNA gave more consistent standard curves and a lower limit of detection (down to 100 copies) while having a similar amplification efficiency as the more laborious RNA dilution method.

13

BIOLUMINESCENT BACTERIA

Whole body biophotonic imaging (BPI) is a technique that has contributed significantly to the way researchers study bacterial pathogens and develop pre-clinical treatments to combat their ensuing infections *in vivo*. Not only does this approach allow disease profiles and drug efficacy studies to be conducted non-destructively in live animals over the entire course of the disease, but in many cases, it enables investigators to observe disease profiles that could otherwise easily be missed using conventional methodologies. The principles of this technique are that bacterial pathogens engineered to express bioluminescence (visible light) can be readily monitored from outside of the living animal using specialized low-light imaging equipment, enabling their movement, expansion and treatment to be seen completely non-invasively. Moreover, because the same group of animals can be imaged at each time-point throughout the study, the overall number of animals used is dramatically reduced, saving lives, time, and money. Also, as each animal acts as its own control over time, the issues associated with animal-to-animal variation are circumvented, thus improving the quality of the biostatistical data generated.

The ability to monitor infections *in vivo* in a longitudinal fashion is especially appealing to assess chronic infections such as those involving implanted devices. Typically, bacteria grow as biofilms on these foreign bodies and are reputably difficult to monitor with conventional methods. Because of the non-destructive and non-invasive nature of BPI, the procedure can be performed repeatedly in the same animal, allowing the biofilm to be studied in situ without detachment or disturbance. This ability not only allows unique patterns of disease relapse to be seen following termination of antibiotic therapy but also *in vivo* resistance development during prolonged treatment, both of which are common occurrences with device-related infections. This chapter describes the bioluminescent engineering of both Gram-positive and Gram-negative bacteria and overviews their use in device-associated infections in several anatomical sites in a variety of animal models.

Historical Overview

In 1995, Contag et al. first demonstrated the ability to monitor bioluminescently engineered bacteria in live mice using whole body biophotonic imaging (BPI). Subsequent to this publication, a range of both Gram-positive and Gram- negative bacteria were bioluminescently transformed and tested *in vivo* in a number of standard animal models, so establishing this technique as a platform methodology for basic research and drug development. Most of these initial studies were focused upon acute models of infection such as soft tissue, sepsis, and pneumonia. It was quickly realized, however, that this technique was especially useful for monitoring chronic and complex infections, such as biofilm infections.

Modern medical and surgical practice has come to rely increasingly on various types of prosthetic and implanted devices. Despite sterilization, aseptic procedures, and improvements in the materials and design of devices, microbial colonization and biofilm formation on these devices is a common occurrence. These biofilm infections represent a serious medical problem, as mature biofilms are resilient to the host immune response and conventional antibiotic treatments. Most often such complications lead to failure of the implant and the need for its complete removal. To better understand and control biofilm infections on indwelling medical devices, new approaches are essential, especially in *in vivo* settings. To this end, BPI methodologies have been developed to allow rapid, continuous real-time monitoring of biofilm infections in a number of different animal models.

The use of bioluminescent reporters to tag bacterial pathogens allows for the sensitive detection of only live, metabolically active cells by BPI. Because of its non-destructive and non-invasive nature, the imaging procedure can be performed repeatedly without disturbing the integrity of the biofilm structure, as is the case with procedure that involves detachment of the bacteria. Additionally, BPI experiments can be performed without the loss of contextual influences of the animal's host defense mechanisms.

Many of the biofilm models described in this chapter resemble characteristics commonly found in human foreign-body infections, such as the establishment of infection with low inoculum, persistence of infection without spontaneous healing, and relapse of disease soon after treatment. This sensitive and quantitative longitudinal monitoring approach is ideally suited to assess disease progression and allows both response and/or relapse to antimicrobial agents to be seen directly from intact live animals in real time, offering an excellent preclinical strategy to assess disease response. Furthermore, this approach is ideally suited to assess novel therapeutic agents or medical devices. This will accelerate efforts to optimize lead compounds, especially at the animal modeling step with respect to formulation, dose, or optimizing materials used in medical implants.

Materials

1. Strains of bacteria: *Staphylococcus aureus* strain ATCC 12600 capsule type 3 strain, *Pseudomonas aeruginosa* strain ATCC 19660, *Proteus mirabilis* strain ATCC 51286, and *Escherichia coli* strain S17-λ A *pir* pUT Amp^R mini-Tn5 *luxCDABE* Tc^R.
2. Plasmid pXen-5.
3. Ampicillin, tetracycline, erythromycin, and kanamycin.
4. $MgSO_4$, glycerol, glucose, Luria-Betani (LB) broth, trypticase soy broth (TSB), and brain heart infusion (BHI) media.
5. Ketamine and xylazine.
6. GenePulser II and cuvettes.
7. Teflon intravenous catheter, 14-gauge.
8. Polyethylene PE 50 (OD 0.965 mm, ID 0.58 mm).
9. IVIS Imaging System and LivingImage software.
10. Ultrasonic bath (VWR).
11. Surgical staples, scissors, and blades.
12. Balb/C, CF-1, and CD-1 female mice weighing 18–30 g.

Methods

Generation of Bioluminescent Bacteria

To maintain bioluminescence stably, all bacteria used for BPI studies were transformed so that an optimized *Photorhabdus luminescense lux* operon is integrated into their chromosome. These operons encode the genes necessary for both luciferase and substrate production, enabling bioluminescence to be produced by the transformants without the addition of exogenous products. Two methodologies for stably transforming the particular Gram-negative and Gram-positive bacteria used in the biofilm studies to a bioluminescent phenotype are described below.

Transformation Procedure for Gram-negative Bacteria

Gram-negative bacteria are made bioluminescent (e.g., *P. aeruginosa* Xen-5 and *P. mirabilis* Xen-44) through the stable integration of an unaltered *P. luminescense luxCDABE* cassette onto their chromosome as first described by Winson et al.

1. Mate the Gram-negative strain to be transformed (the recipient) with the *E. coli* strain S17-λ *pir* pUT Amp^R mini-Tn5 *luxCDABE* Tc^R(the donor).
2. To allow the recipient strain to be more readily distinguished from the donor strain, initially transform the recipient with a blank vector carrying a unique antibiotic resistance gene (primary resistance marker) using standard electroporation procedures.
3. Use a single resistant transformant (carrying the primary resistance marker) to inoculate 10 mL of LB containing the primary selective antibiotic and incubate overnight at 37°C. Similarly, use a single colony of *E. coli* S17-λ *pir* pUT mini-Tn5 *luxCDABE* Tc^R to inoculate 10 mL of LB containing 100 μg/mL ampicillin.
4. Individually, pellet and resuspend both bacterial cultures in 10 mL of LB.
5. Inoculate a fresh 10-mL volume of LB containing 10 m*M* $MgSO_4$ with 100 μL volumes of each of the suspensions and incubate this mixture overnight at 37°C to allow mating to occur.
6. Plate 100 μL volumes of the mating onto LB containing both the primary selective antibiotic and 20 μg/mL tetracycline.
7. After an overnight incubation at 37°C, screen the plates for bioluminescent transposants using a highly sensitive optical imaging system, such as an IVIS Imaging System.
8. Select a single highly bioluminescent colony and characterize to ensure that it is stable [e.g., bioluminescent photon counts correspond to colony-forming units (CFUs)] and behaves similarly to the parental strain, both *in vitro* and *in vivo* (e.g., not attenuated).

Transformation Procedure for Gram-positive Bacteria (*Staphylococcus aureus* Methodology)

Gram-positive bacteria are made bioluminescent (e.g., *S. aureus* Xen-29) through the stable integration of an optimized *P. luminescense luxABCDE* cassette onto their chromosome as first described by Francis et al. This broad host range plasmid can be used to transform a variety of Gram-positive bacteria to a stable bioluminescent phenotype. However, the procedure described below is optimized for the transformation of *S. aureus*.

1. Grow-up an overnight culture of *S. aureus* in 10 mL of BHI.
2. Dilute the overnight culture 1:100 in 50 mL of fresh BHI broth in a 500-mL flask and incubate at 37°C in an orbital shaker at 200 rpm until the OD_{600} reached 0.8 (exponential growth phase).
3. Centrifuge and then sequentially wash cells with 50 mL, 25 mL and 5 mL ice-cold 10% glycerol solution (centrifuging between washes).
4. Resuspend the electrocompetent cells in 500 μL of ice-cold 10% glycerol.
5. Add one 1 μL of 1 μg/μL plasmid pXen-5 (pAUL-A Tn*4001 luxABCDE* Km^R) to 200 μL of electrocompetent *S. aureus* and mix.
6. Transfer the bacteria/plasmid mixture into a 1-cm ice-cold cuvette and electroporate at 25 μF, 100 Ω, and 2.5 kV.
7. Immediately add 0.8 mL of BHI, gently invert cuvette and incubate statically at 37°C for 1h.
8. Plate 200 μL volumes of the transformation mix on BHI agar plates containing 0.3 μg/mL of erythromycin and incubate plates at 37°C for 24 h.
9. Patch colonies (~1 cm^2) onto BHI agar plates containing 0.3 μg/mL of erythromycin and incubated at 37°C overnight.
10. Uniformly streak a quantity of each patch (10 μL loopful of cell growth consisting of ~10^8–10^9 cells) over the entire area of a BHI agar plate containing 200 μg/mL of kanamycin and incubated plates at 37°C for 24 h.
11. Screen plates for bioluminescent transposants using a highly sensitive low-light optical imaging system.
12. Select a single highly bioluminescent colony and characterize to ensure that it is stable (e.g., bioluminescent photon counts correspond to CFUs) and behaves similarly to the parental strain, both *in vitro* and *in vivo* (e.g., not attenuated).

In vitro Catheter-Associated Biofilm

1. Grow the bioluminescent biofilm-forming bacterial strain overnight in TSB supplemented with 0.25% glucose (TSBG) in an orbital shaker at 37°C.

2. Dilute the culture 1:10 in fresh TSBG and further incubate for 1.5 h at 37°C.
3. Adjust OD at 600 nm to reach a standardized cell suspension of 10^5 CFUs/mL in TSBG.
4. Cut teflon intravenous catheter 14-gauge into 1-cm segments (soft tissue model), polyethylene catheter PE 50 into 6-mm segments (UTI) and sterilize each piece with 70% ethanol. Air-dry catheters overnight.
5. Develop bacterial biofilms on the cut, sterile catheters by placing individual segments into Eppendorf tubes containing 1.0 mL of bacterial cell suspension in TSBG with 10^5 CFUs/mL in the exponential phase of growth.
6. Vortex well to ensure the removal of air bubbles from inside catheter, allowing it to remain immersed during incubation.
7. After incubating for appropriate time intervals at 37°C recover colonized catheters aseptically and rinse once gently in TSBG to remove unbound bacteria. Flush out lumen of catheter with sterile TSBG. Image catheters for bioluminescence prior to implant to ensure proper colonization.

Soft-tissue Biofilm Infection

This device-related infection is established in the flanks of mice by subcutaneous implantation of either pre- or *in vivo*-colonized catheter material with biofilm forming Gram-positive or Gram-negative bioluminescent bacteria. The model is ideally suited for screening the development of biofilms on medical device material and the effects of therapeutic agents to treat them non-invasively without the need for exogenous sampling or culturing of the pathogen. Infections could be established with clinically relevant doses as low as 10^3 CFUs pathogen per device with either Gram-positive or Gram-negative pathogens. Moreover, subsequent disease progression could be followed for weeks within the same animal.

This model is especially appealing for the analysis of *in vivo* efficacy of antibiotics against young as well as mature biofilms, post antibiotic effects, *in vivo* resistance development, or *in vivo* fitness of resistant organisms. As the imaging procedure can be repeatedly performed over time without killing the animal or removal of device for detachment of biofilm, this model, with several of the clinically relevant features, has proven to have significant advantages over conventional methods for studying foreign-body-related infections and treatment *in vivo*.

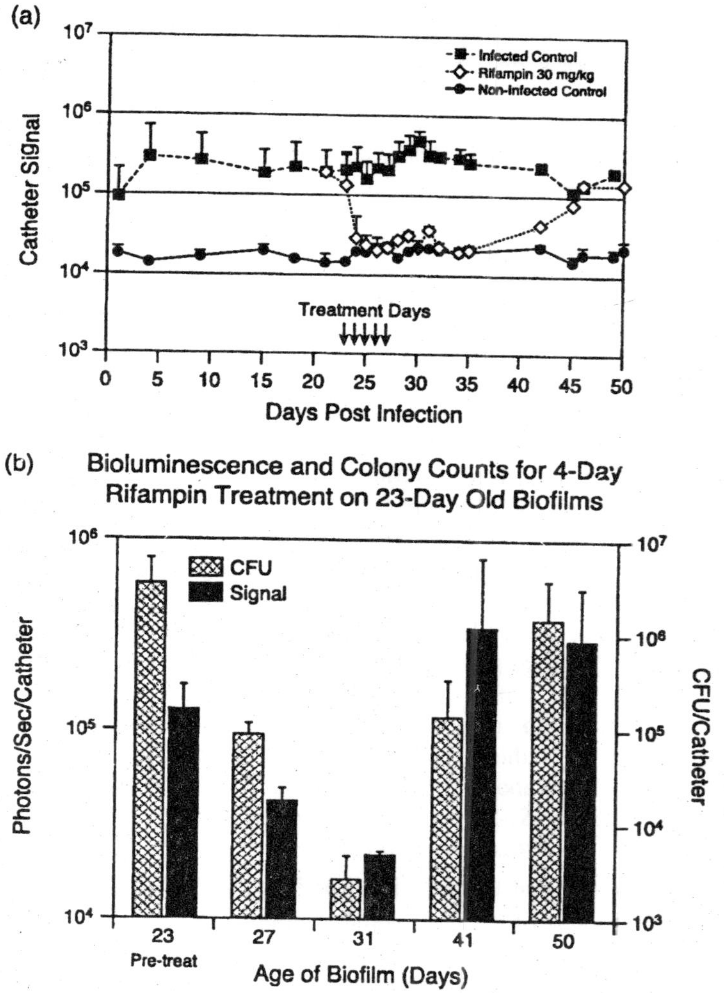

Fig. 13.1. A–In vivo bioluminescence monitoring of S. aureus Xen29 in the mouse model of biofilm infection during treatment with rifampin. Effect of 4 days of rifampin treatment on a 23-day-old biofilm. B–Correlation between bioluminescence and traditional colony-forming units (CFUs) for 4-day rifampin treatment on a 23-day old biofilm.

Experimental Model of Infection

1. Anesthetize Balb/C female mice weighing 18–22 g with Ketamine (100 mg/kg) and Xylazine (5 mg/kg).
2. Shave their flanks and clean the skin with Betadyne and alcohol.

3. Make a 4–5 mm skin incision (as small as possible) and dissect to create a subcutaneous tunnel.
4. Push a 1-cm segment of intravenous catheter with its associated bioluminescent bacterial biofilm through the incision into the tunnel (subcutaneous tunnel usually created is ~1.5 cm in length). One catheter segment is inserted on each side of each animal.
5. Cover the incisions with intact skin, close with surgical staples, and disinfect the skin.
6. Following surgery, place the mice on a warming pad until fully recovered.
7. Image animals using a highly sensitive low-light optical imaging system
8. If post-implant infection is preferred, implant segments of sterile catheters as above, then approximately 1 h after the implantation procedure, introduce a defined quantities of bacterial suspension in 50 μL of PBS by injection into the catheter lumen (10^3–10^5 CFUs/catheter) via a 31-gauge needle.

Delaying the infection until after device implantation is useful to study factors such as post-operative colonization of implanted device or the role of host proteins in promoting bacterial adhesion to biomaterials in initial phase of blood–material interaction. An inoculum of ~10^3–10^5 CFUs of *S. aureus* Xen-29 and *P. aeruginosa* Xen-5 per catheter results in a reproducible, localized persistent infection surrounding the catheter until the termination of the experiment on day 25–50. Doses above 10^6 CFUs/catheter for *P. aeruginosa* and 10^8 CFUs/catheter for *S. aureus* result in 100% mortality. An inoculum of $\leq 10^5$/catheter produces chronic infections.

Catheter-Associated Urinary Tract Infections

UTIs are one of the most common bacterial infections in humans. Most of these infections follow instrumentation of the urinary tract, mainly urinary catheterization, with the development of catheter-associated bacteriuria directly related to the duration of catheterization. BPI easily allows spatial information to be monitored sequentially throughout the entire disease process, from cystitis to ascending UTIs, as well as treatment efficacy and relapse in diseased or asymptomatic animals all without exogenous sampling.

Experimental Model of Infection

The experimental catheter-associated UTI in a rat model developed by Kurosaka et al. was established in mice with modifications for BPI.

1. Cut polyethylene catheter material PE 50 into 6-mm long segments and fit onto a flexible metal wire.
2. Wrap this around a similar metal wire in a spiral and place in boiling water for 1 min.
3. Remove the segment and sterilize each piece with 70% ethanol and air dry overnight.
4. Bacterial biofilm is grown on the catheter material as described above.
5. Cut the metal stylet of a 24-gauge Teflon catheter up to 5 cm and fit either sterile PE tubing or tubing that has been colonized with bacteria.
6. Anesthetize CF-1 female mice weighing 26–30 g with 2.5% isoflurane gas and place in the supine position.
7. Clean the periurethral area with 70% ethanol.
8. Grasp the papilla with light pressure using forceps and lift away from the abdomen.
9. While maintaining the papilla in this position, place the end of the stylet bearing the 6-mm-long tubing into the urethral opening and advance until the leading end is in the bladder.
10. Push the 5-cm Teflon catheter until the shorter segment is pushed off the stylet, thereby leaving the 6-mm-long PE tubing free in the bladder lumen (the PE tubing will revert spontaneously to its spiral form thus remain in place until the end of the monitoring period).
11. Remove the Teflon catheter with the stylet.
12. Image animals using a highly sensitive low-light optical imaging system.
13. Remove Teflon catheter post-inoculation.

To induce post-catheterize infection, place a sterile tubing in the bladder as described above and inoculate the bladder with defined quantity of pathogen in 50 μL phosphate-buffered saline over 30 s by insertion of Teflon catheter into the bladder through the urethra.

Central Venous Catheter-Associated Infection in Mice

Long-term implantation of intravascular devices is one of the major advances in modern medical and surgical practice. A common complication with these implanted foreign bodies is their susceptibility to pyogenic infections, especially due to *Staphylococci*. Establishment of such infections is due to either perioperative bacterial contamination or postoperative colonization of the implanted device through local or hematogenous seeding. Little is known about the pathogenesis and treatment of this condition, mainly due to lack of in situ methods to monitor such infections *in vivo* in real time. Thus, we have applied the *in vivo* imaging technology to monitor and visualize the establishment and spread of staphylococcal infection non-invasively in mice with indwelling jugular vein catheter .

Experimental Model of Infection

To avoid high operative mortality rate associated with small animal vascular catheterization, it is advisable to purchase animals that are surgically prepared and ready to use from a commercial vendor such as Charles River Laboratories.

1. After 5–7 days post-catheterization, infect the catheterized mice intravenously with 50 μL of bioluminescent bacterial suspension (e.g., *S. aureus* Xen 29) in exponential growth phase in PBS through the tail vein with inocula of 10^4CFUs.
2. Inoculate a group of non-catheterized mice with a similar dose as control group.
3. Image animals using a low-light imaging system.
4. To optimize capture of bioluminescence, shave over the heart (ventral), as well as kidneys and spleen areas (dorsal).

Separation of Bacteria From Catheter

1. After the final imaging time-point, kill mice humanely and surgically remove the infected catheters for enumeration of bacteria by both bioluminescence imaging and conventional plate count method.
2. Image the harvested catheters for bioluminescence.
3. Transfer the catheters to tubes containing 1 mL of TSB, place tubes in an ultrasonic bath (VWR), and sonicate for 5–10 min (38.5–40.5 kHz).
4. Vortex the tubes for 1 min to remove the biofilm bacteria from the support surface and image catheters to confirm the loss of bioluminescence and thus the complete removal of the biofilm.
5. Repeat sonicating-vortexing procedure if necessary. Note, repeating the procedure for >10 min may affect the viability of certain pathogens.
6. Dilute the suspension of bacteria removed form the catheter, plate on trypticase soy agar, and incubate at 37°C for colony counting.

Correlation between CFUs and bioluminescent signal is determined by plotting Photons/Sec/Catheter versus CFUs/Catheter.

Imaging

1. Prior to and during imaging, mice are anesthetized with 2–2.5% isoflurane gas.
2. The whole animal is imaged (dorsal and ventral) for a maximum of 5 min at various times during the experiment using a highly sensitive optical imaging system, such as an IVIS Imaging System.
3. Quantify total photon emission from defined regions of interest (e.g., bladder, kidney, flanks, abdomen, or heart) within the images of each animal.
4. Following imaging, return mice to their cages until the next imaging session.

Notes

1. pUT is a conjugative suicide vector that is only maintained in strains of *E. coli* expressing the π protein, such as S17-λ *pir* and CC118 λ *pir*. Thus, recipient bacteria that do not express the π protein can only become stably bioluminescent and tetracycline resistant if the mini-Tn5 *luxCDABE* Tc^R cassette transposes onto their chromosome. Furthermore, as the *lux* operon does not have its own promoter, the strength of bioluminescence in the recipient bacterium is dependent upon the endogenous promoter directly upstream of the site at which the *lux* transposon integrates. Derivatives of this donor strain carrying alternative antibiotic markers are also available (e.g., chloramphenicol, kanamycin, and streptomycin).
2. The promoterless *luxABCDE* km^R operon is optimized from Gram-positive expression and is silent on pXen-5. When transposition occurs, the strength of bioluminescence and the degree of kanamycin resistance in the bacterium is dependent upon the endogenous promoter directly upstream of the site at which the *lux* transposon integrates.
3. The following conversion factors are useful in adjusting cell suspensions to the desired levels: *S. aureus* Xen 29 OD_{600} 1.0 = 2.8 × 10^8 CFUs/mL, *P. aeruginosa* Xen 5 OD_{600} 1.0 = 7.5 × 10^8, and *P. mirabilis* Xen 44 OD_{600} 1.0 = 5.4 × 10^8 CFUs/mL.
4. Incubation time for *S. aureus* Xen 29 to reach 10^3 CFUs/Catheter = 30 min, 10^4 CFUs/Catheter = 1 h, and 10^5 CFUs/Catheter = 2h in 10^5 CFUs/mL of suspension. For *P. aeruginosa* Xen 5 to reach 10^3 CFUs/Catheter = 10 min, 10^4 CFUs/Catheter = 1.5 h, and 10^5 CFUs/Catheter = 3 h in 10^5 CFUs/mL of suspension. The

above values are valid for 14-gauge, 1-cm segments of Teflon intravenous catheter. The concentration of cell suspension and time required to reach desired inoculum varies with surface area and the type of material.

5. The challenge doses required to establish chronic UTI with *P. aeruginosa* Xen 5 is 1.6×10^6 CFUs/catheter or 2.0×10^6 CFUs/50-μL of suspension and with *P. mirabilis* Xen 44,7.1×10^3 CFUs/catheter or 1.3×10^4 CFUs/50-μL of suspension.

14

METHYL HALIDES DEGRADATION

The monohalomethanes methyl chloride (or chloromethane; MeCl), methyl bromide (or bromomethane; MeBr), and methyl iodide (or iodomethane; MeI) have received significant attention over recent years. Due to its high ozone depleting potential, the use of MeBr in industrial and agricultural applications, for example, soil fumigation, is now being phased out under the Montreal Protocol. Concerns about the damaging effects of methyl halides have sparked investigations into the sources and sinks of these compounds. This has led to a better appreciation of the natural cycle of methyl halides. Degradation of methyl halides by microorganisms is recognized as a major sink for these compounds in the environment and a number of novel microorganisms involved in methyl halide degradation have been isolated. Studies of the metabolic pathways and the biochemistry and genetics of bacterial methyl halide degradation have given some insight into this important biological activity. This chapter focuses on the monohalogenated methanes MeCl and MeBr, their natural and anthropogenic sources, and their degradation by microorganisms, specifically by aerobic bacteria that can use MeBr and MeCl as sole source of carbon and energy. The biogeochemical cycle of methyl halides, and the microbiology, biochemistry, genetics, and biotechnological potential of methyl halide degrading microorganisms are discussed.

HISTOLOGY

Role of Methyl Halides in Atmospheric Chemistry and as Ozone Depleting Compounds

Methyl halides are the dominant halocarbons in the atmosphere. As such, they play an important role in regulating stratospheric ozone concentrations and to some extent, global warming as well, two factors governing planetary habitability. The monohalomethanes considered here, methyl chloride (MeCl), methyl bromide (MeBr), and methyl iodide (MeI), are trace gases in the atmosphere with average tropospheric mixing ratios of 600, 10, and 2 parts per trillion (ppt), respectively. However, methyl halides are radiatively active and hence contribute to global warming by absorbing radiation in the infrared region. This is evident in their elevated global warming potential (GWP), a value calculated to quantify each compound's warming effect (on a mass basis) relative to the same mass of CO_2. Compounds with high GWP, including those like methyl halides with low concentrations, may have considerable impact on atmospheric warming when compared with other "greenhouse" gases with low GWP.

More importantly, methyl halides are the major natural sources of halogen-containing radicals to the stratosphere (altitude >15 km) where they participate in the catalytic destruction of ozone. These reactions contribute to the "ozone holes" which have appeared during springtime in the upper atmosphere above polar regions since the 1980s. Methyl halides are also partly responsible for the gradual thinning of the effective layer of ozone worldwide, leading to greater penetration of ultraviolet (particularly UV-B) radiation. It is estimated that MeBr contributes more than half of the Br present in the stratosphere and that MeCl is responsible for at least 15% of chlorine-catalyzed ozone consumption. This is reflected in their elevated ozone depletion potential (ODP) which is calculated to indicate their ability to destroy ozone in the stratosphere, again on a mass basis relative to the same mass of the chlorofluorocarbon CFC-11 (CCl_3F).

Ozone-depleting compounds, including CFCs and MeBr, have been regulated since 1987 by the Montreal Protocol. This international agreement encourages signatory nations to decrease emissions of all ozone depleting compounds and has effectively reduced the burden of chlorine and bromine in the atmosphere. CFCs, halons (brominated compounds used as fire retardants), and other synthetic compounds were the first to be phased out and their concentrations in the atmosphere are decreasing. MeBr concentrations have also decreased, but the matter is greatly complicated by the fact that only about 30% of the MeBr released is man made and therefore subject to direct control.

Biogeochemical Cycle of Monohalomethanes

The majority of methyl halides are produced naturally at the Earth's surface. Of major concern is the increased concentration and flux of atmospheric methyl halides (particularly MeBr) over the past 50–100 years, resulting from changes in the magnitude of sources and sinks. MeBr and MeCl react with atmospheric hydroxyl (OH) radicals and this oxidation represents their primary removal mechanism from the troposphere. The balance between release at the Earth's surface and loss in the troposphere results in sufficient atmospheric lifetimes such that significant quantities of these compounds are able to reach the stratosphere where they can react with ozone. By contrast, MeI is removed more rapidly by photolysis in the troposphere and its shorter lifetime limits its impact on stratospheric ozone. In part this is due to the chemical reactivity of methyl halides, which increases with the atomic number of the halide ion (e.g., MeF < MeCl < MeBr < MeI). Uncertainty, however, clouds our understanding of the budgets of methyl halides. Currently, known sinks of MeBr and MeCl outweigh sources by about 35%. The budget for MeI is balanced, although there are large uncertainties in estimates of the oceanic source and the photolysis sink.

Anthropogenic Sources

Methyl halides are manufactured for a variety of purposes by reacting acids, containing halogens derived from salt, with methanol from natural gas. Anthropogenic MeBr is used as a preplant and postharvest fumigant for crops in much of the world today. It is further used to fumigate structures for pest control and for quarantine fumigation of commodities shipped worldwide. In addition, MeBr can be used in treating buildings or vehicles contaminated with spores of Bacillus anthracis as a possible counterterrorism measure, although it was commonly used as far back as the 1950s to decontaminate animal hides prior to their handling by tanners. MeI is beginning to replace MeBr as a soil fumigant because it is less hazardous to stratospheric ozone. MeCl and MeBr are used industrially to manufacture other chemicals but little of this is released to the atmosphere. The largest strictly anthropogenic emissions of methyl halides derive from combustion of fossil fuels, particularly coal for MeCl and leaded gasoline for MeBr. These sources are generally comparable to or smaller than atmospheric inputs from biomass burning.

Natural Chemical Sources and Sinks

A fraction of methyl halides are created or destroyed naturally by chemical processes. This fraction may be large, as in the case of MeI production by photochemical reactions in surface seawater. Similarly, a large quantity of MeCl is produced in soils by reaction of plant derived pectin with inorganic Cl^-. A significant but unknown quantity of methyl halides is produced by the oxidation of soil organic matter in the presence of ferric iron and halides in soil and sediments. In aqueous environments, methyl halides are transformed by nucleophilic substitution of one halide for another, known as halide exchange. Hydrolysis, a similar chemical process involving nucleophilic substitution with water or hydroxide ion (OH^-), results in the removal of methyl halide and the production of methanol. The impact of these abiotic reactions remains unquantified, but they are of importance in the balance of methyl halides in soil and seawater.

Natural Biological Sources

By far, the largest sources of methyl halides to the atmosphere are the result of naturally occurring biological activity on the Earth's surface. This review examines the role that biology plays in affecting the balance between production and consumption of methyl halides in aquatic and terrestrial environments. Methyl halides are produced by a variety of cyanobacteria and marine algae, including phytoplankton and macroalgae, as well as some terrestrial plants and fungi. In algae, higher plants, and fungi, methylation of inorganic Br^- or Cl^- by S adenosyl L methionine (SAM) to form S adenosylhomocysteine and MeBr or MeCl is catalyzed by methyltransferase enzymes. Methyl halides are produced in large quantities by halophilic plants growing in salt marshes. However, this production is not particularly effective at removing excess halides. Rather, methyl halide production may simply be a metabolic coincidence resulting from the proximity of halide ions and methyltransferase enzymes within the plant cell. Production of methyl halides may provide a competitive advantage to some microorganisms, particularly fungi, through direct antimicrobial effects on their competitors/predators or by providing substrate for the biosynthesis of secondary metabolites which facilitate lignin degradation.

Natural Biological Sinks

Consumption of methyl halides is, likewise, largely under the control of biological processes occurring on the Earth's surface. In anaerobic sediments containing free sulfide, MeBr and MeCl may be transformed chemically to methanethiol (CH_3SH) or dimethylsulfide [$(CH_3)_2S$], which are substrates for both sulfate reducing bacteria and methanogenic archaea. However, few anaerobic zones are in direct contact with the atmosphere at the Earth's surface, hence the largest biological sinks for atmospheric methyl halides exist in aerobic environments. Lobert et al. (1995) demonstrated that the oceans were a net sink for atmospheric MeBr and subsequent work by Pilinis et al.

(1996) suggested that biological activity in the oceans was responsible for some of the uptake. Shorter et al. (1995), working with a diversity of live soils, used flux measurements and additions of antibiotics to demonstrate that globally scaled MeBr consumption from the troposphere could be attributed to the metabolic activities of aerobic bacteria living within the tested soils. These results were significant in that they demonstrated a clear biological consumption of realistic tropospheric-mixing levels of MeBr, which are about 10–12 ppt. No other previous research on microbial trace gas consumption had dealt with microbial metabolism at such low concentrations. The ongoing assumption was that the affinity of oxidative enzymes for such rare trace gases would not be operative below some threshold boundary of say 100 ppb.

Stable Isotope Mass Balances and Fractionation

It was previously mentioned that large uncertainties exist in the magnitude of the terms comprising the global budgets of methyl halides. These uncertainties may be reduced by constructing stable carbon isotope mass balances. McCauley et al. (1999) first suggested constraining the atmospheric budgets of MeBr by analyzing the carbon isotopic composition of major sources and the fractionation associated with sinks (i.e., removal mechanisms). They determined the $\delta^{13}C$ value for industrial MeBr (–54.4%o vs Vienna Peedee Belemnite; VPDB). This group and others determined the isotopic composition of MeBr entering the atmosphere from salt marshes and soil fumigations (–43‰ and –48.7‰, respectively). They further measured the isotopic composition of MeBr in the atmosphere (–44.4‰ to –41.2‰), which allows closer examination of the effect of sinks on the stable isotope mass balance of methyl halides.

Similar progress has been made with global MeCl mass balances. MeCl derived from biomass burning is released to the atmosphere with a $\delta^{13}C$ value about 25‰ depleted from the starting plant material or about –47‰ ± 12‰ versus VPDB. Tropical vegetation is a large source of MeCl with low $\delta^{13}C$ value (–72.7‰ to –69.3‰). Isotopically ^{13}C-depleted MeCl is also emitted by salt marsh plants (–62‰) and wood-rotting fungi (–43‰). Keppler et al. (2004) reported that MeCl released by low-temperature heating of plant leaves was highly depleted in ^{13}C with values ranging from –147‰ to –119‰. The source of MeCl was the reaction of plant pectin (a ^{13}C-depleted fraction of the methoxyl pool) with chloride ion. The stable C isotopic composition of MeCl in the atmosphere ranges from –44‰ to –30‰ with a global average of –36.2 ± 0.3‰.

Large C isotopic fractionations (up to 72‰ for MeBr and up to 50‰ for MeCl) have been observed for bacterial reactions that oxidize methyl halides. Large discrimination in favor of ^{12}C over ^{13}C was observed during oxidation of methyl halides by pure cultures of *Aminobacter ciceronei* IMB-1 and *Leisingera methylohalidivorans* MB2. However, for unknown biochemical reasons, MeBr degradation by *Aminobacter lissarensis* CC495 proceeded without such C fractionation. Patterns of stable C isotope measurements over time were part of a suite of tools used to distinguish biological from chemical removal processes in soil microcosms. In water, chemical removal by hydrolysis or halide exchange reactions results in large fractionation of carbon isotopes (33–57‰), resulting in enrichment of ^{13}C in the remaining methyl halide. These sinks may be significant in soil and fresh water at the Earth's surface and especially in the ocean, however, the impact these sinks have on the isotopic composition of methyl halides released to the atmosphere requires further investigation.

The largest sink for atmospheric MeBr and MeCl is the reaction with OH in the troposphere, a process which removes much of the surface flux of these compounds. Studied in the laboratory, these reactions produce large isotope fractionations (^{13}C depletion) by as much as 59 ± 8‰ for MeCl and 66 ± 30‰ for MeBr. These observations reveal a conundrum. If the tropospheric sink terms are large, and these reactions significantly fractionate the remaining unreacted methyl halide pool by enriching it in ^{13}C, then the global average isotope values for atmospheric methyl halides should not be similar to the weighted means of all sources. Put as a question, why is there little effect of the combined sink terms on the isotopic composition of the unreacted methyl halides in the troposphere? Could it mean that there are significant unidentified sources of isotopically "light" methyl halides to the atmosphere? Or perhaps the large fractionation attributed to OH radical oxidation is valid for laboratory settings only, and does not apply to the chemically more complex troposphere? These questions must be addressed before isotope mass balances can be successfully used to constrain global budgets, but it nonetheless illustrates the utility of stable C isotope balances in shaping testable hypotheses.

Methyl Halide-degrading Organisms

Bacterial Degradation of Methyl Halides by Methanotrophs and Nitrifiers

A wide variety of microorganisms capable of degrading methyl halides have been described. First reports of bacterial methyl halide degradation were based on observations that certain bacteria were able to co-oxidize, but not grow on methyl halides. Methanotrophs such as *Methylococcus capsulatus* (Bath), *Methylosinus trichosporium*, and *Methylomonas methanica* were able to co-oxidize MeCl using the enzyme methane monooxygenase even though they could not grow on methyl halides as a carbon source. It has also been shown that metabolism of MeCl enhanced growth of *Methylomicrobium album* BG8 on methanol. Conversion of MeCl to carbon dioxide occurred, suggesting that co-oxidation of methyl halides by particulate methane monooxygenase may provide metabolic intermediates that can be assimilated by methanotrophs. Ammonia-oxidizing bacteria were also shown to degrade methyl halides. Using a bromide selective electrode, Hyman and Wood demonstrated that washed cell suspensions of *Nitrosomonas europaea* were able to generate bromide ions from MeBr at rates that were much higher than rates of chemical hydrolysis of MeBr. The generation of bromide slowed down to chemical hydrolysis rates when the medium became anoxic. Rasche and colleagues showed by monitoring headspace gas concentrations that *N. europaea*, *Nitrosococcus oceani*, and *Nitrosolobus multiformis* degraded MeBr. Inhibition of MeBr degradation in these strains by allylthiourea, a specific inhibitor of ammonia monooxygenase (AMO), suggested that AMO was the enzyme responsible for MeCl degradation.

Diversity and Distribution of Bacteria Capable of Growth on Methyl Halides as a Carbon and Energy Source

MeCl and MeBr have both been used successfully in enrichment experiments resulting in the isolation of a variety of bacterial cultures that can grow on methyl halides. Methyl halide-degrading bacteria have been isolated from a range of environments, including agricultural, woodland and industrially contaminated soils, activated sludge, lake sediments, estuarine and seawater samples, suggesting that methyl halide-degrading bacteria are ubiquitous in the environment. Bacteria growing on methyl halides are mostly members of the *Alpha Proteobacteria*, but also of the *Gamma Proteobacteria* and the *Actinobacteria*.

Isolates from activated sludge

Bacterial growth on MeCl as a sole carbon source was first reported for *Hyphomicrobium* strain MC1 which was isolated from activated sludge. Unfortunately this strain has been lost. Activated sludge was also the source of strain MC, a strictly anaerobic Gram positive, homoacetogenic bacterium, which grew on MeCl as sole source of carbon and energy.

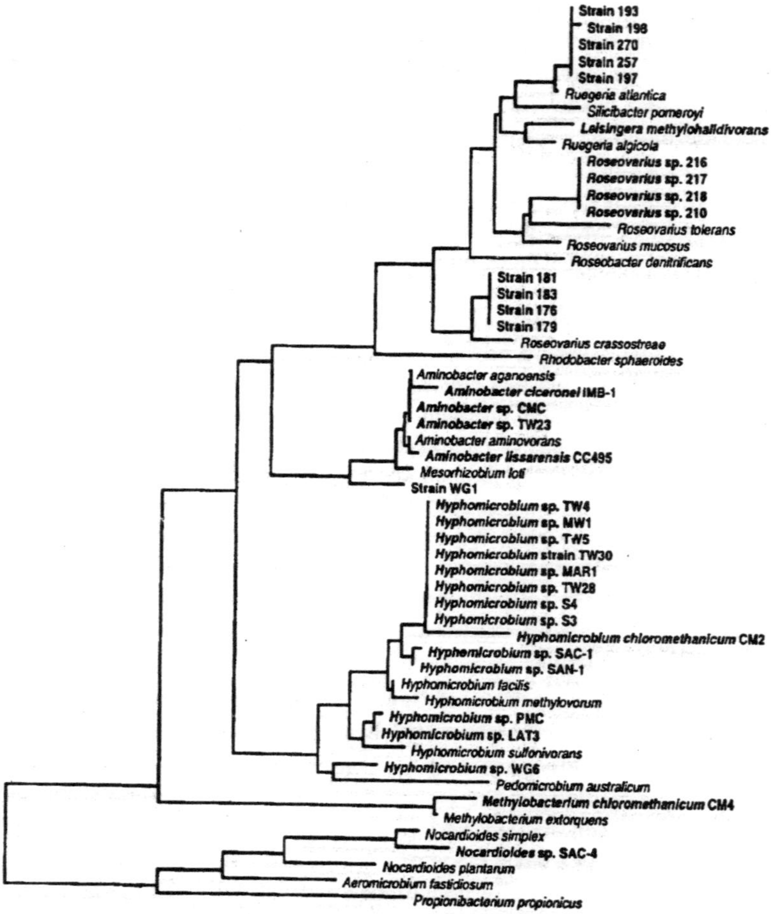

Fig. 14.1. Neighbor joining tree based on partial 16S rRNA gene sequences showing the diversity of bacteria capable of utilizing methyl halide as a growth substrate.

Soil and freshwater isolates

Doronina et al. (1996) isolated *Hyphomicrobium* and *Methylobacterium* strains from contaminated soil at the site of a petrochemical factory in Russia. These strains assimilated formaldehyde using the serine cycle and were able to grow on MeCl as sole source of carbon and energy using an unknown inducible enzyme system for degradation of MeCl.

Subsequently further facultative methylotrophs capable of methyl halide degradation were isolated from the terrestrial environment. Strain

IMB 1 was obtained from a southern Californian strawberry field soil that had been subject to fumigation with MeBr, and was initially identified as being closely related to members of the genus *Rhizobium*. It was isolated by using MeBr as its sole source of carbon and energy. The high toxicity of MeBr was circumvented by making a series of pulsed, low concentration additions of MeBr to the gas phase (typically ~0.5%) over the course of the incubations. A similar strategy was used to demonstrate growth of this isolate on MeI. A related isolate that grew on MeCl, strain CC495 was obtained from a woodland soil in Northern Ireland by Coulter et al. (1999). These isolates have been characterized as novel species *A. ciceronei* IMB 1 and *A. lissarensis* CC495. Additional strains of *Hyphomicrobium* were isolated from a variety of woodland soils and also from lake sediments. The only aerobic Gram positive strain growing on MeCl that has been described was a *Nocardioides* strain (strain SAC-4), isolated from a woodland soil. Unfortunately this isolate has been lost. An isolate of *Pseudomonas aeruginosa*, strain NB1, was shown to degrade MeCl under aerobic and denitrifying conditions.

Marine methyl halide utilizers

Despite the importance of the marine environment as a sink in the global cycle of methyl halides, only a few marine strains growing on methyl halides have been isolated. Although a *Hyphomicrobium* strain was isolated from the Severn estuary in the UK, *Hyphomicrobium* spp. are not typical marine bacteria and so the isolation of *L. methylohalidivorans* MB2 from a tidal pool off the Californian coast can be considered the first marine strain growing on MeBr. Strain MB2 was characterized as a facultative methylotroph belonging to the *Roseobacter* clade, an abundant group of microorganisms in seawater. Subsequently a further 13 facultative methylotrophic strains growing on MeBr, affiliated with the *Roseobacter* clade, were isolated from the Scottish coast and the English Channel. These strains belonged to three different species of the *Roseobacter* clade and were most closely related to members of the genera *Roseovarius* and *Ruegeria*.

Further MeBr-degrading strains were obtained from seawater samples but were not identified by 16S rRNA sequencing. One of these isolates, strain LIS-3, grew poorly on MeBr, another isolate, strain FV co-oxidized MeBr and could not grow on MeBr. Similarly, a strain related to the Sphingomonadaceae, strain Oxy6, cometabolized MeBr. Strain Oxy6 was obtained from a toluene enrichment and MeBr consumption appeared to be cometabolic and inhibited by toluene.

Biochemistry and Genetics of Methyl Halide Degradation

Metabolism of Methyl Halides by Bacterial Isolates

The work of Hartmans et al. (1986) on the first isolate that grew on MeCl, *Hyphomicrobium* MC1, showed that MeCl degradation in strain MC1 released chloride ions in a stoichiometric manner. Methane was not oxidized by the strain, which excluded the possibility that methane monooxygenase was responsible for MeCl degradation. Resting cells of strain MC1 grown on MeCl dechlorinated MeCl with concomitant oxygen consumption, whereas methanol-grown cells did not. Methanol-grown cells had high rates of methanol oxidation, while MeCl-grown cells had low methanol oxidation rates. It was suggested that formaldehyde was an intermediate in MeCl metabolism and that a monooxygenase could be involved. The mechanism of MeCl dehalogenation however was not identified in this strain and since the strain was lost, further biochemical analysis was not possible.

As with strain MC1, methyl halide metabolism generally appears to be inducible in methylotrophs growing on these compounds. This trait has been described for several strains, including *Methylobacterium chloromethanicum* CM4, *Hyphomicrobium chloromethanicum* CM2, and the marine isolates *Roseovarius* sp. 217, *Ruegeria* sp. 198, and strain 179. Although cells of *A. ciceronei* IMB-1 grown in the absence of MeBr had constitutive MeBr degradation activity at low concentrations of MeBr (<20 nM), induction by MeBr was required for consumption of higher MeBr concentrations (100 μM and above). MeBr-oxidation experiments with *L. methylohalidivorans* MB2 in the presence of the protein synthesis inhibitor chloramphenicol indicated rapid turnover of MeBr-degrading enzymes, even when exposed to tropospheric levels (12 ppt) of MeBr.

However, although for *A. ciceronei* IMB-1, the identity of the enzyme system responsible for MeBr consumption at tropospheric-mixing ratios has not been established, it is possible that the enzyme system could be identical to the enzymes responsible for consumption of higher MeBr concentrations. It is interesting to note that the kinetic parameters of MeBr consumption estimated at mixing ratios near tropospheric levels were similar to those derived from experiments using higher mixing ratios and batch incubations. Although K_s values were 10^2–10^3 times higher than tropospheric-mixing ratios, *A. ciceronei* IMB-1 and *L. methylohalidivorans* MB2 both could consume tropospheric levels of MeBr.

The CmuA Pathway of Methyl Halide Degradation in *M. chloromethanicum* Strain CM4

A combination of elegant genetic and biochemical experiments led to the identification of the pathway of MeCl metabolism in *M. chloromethanicum* CM4. Leisinger and coworkers introduced random mutations in *M. chloromethanicum* CM4 by transposon mutagenesis and screened for mutants which could not grow on MeCl, but which could still grow on methanol. Analysis of the site of transposon insertions that lead to a loss of MeCl degradation revealed a number of genes that were (potentially) essential for growth on MeCl as a carbon source. These included genes designated cmuA, cmuB, and cmuC that were predicted to encode methyltransferases, based on similarity shared with methyltransferases from methanogenic archaea. The gene cmuA encoded an unusual protein of 617 amino acids, which appeared to have two functionally distinct domains. While the N-terminal domain of CmuA had similarity to the coenzyme M methyltransferases (MtbA) involved in methanogenesis from methylamine in *Methanosarcina barkeri*, the C-terminal domain was similar to the monomethylamine corrinoid protein (MtmC) of the same organism, which accepts the methyl group from methylamine during growth on this compound. Both CmuA and CmuB were purified from M. chloromethanicum strain CM4 and it was shown in vitro that CmuA and CmuB in combination were able to transfer the methyl group of MeCl to tetrahydrofolate (H_4F). It was also shown that tetrahydromethanopterin did not accept the methyl group from CmuB *in vitro*. Other genes in the vicinity of cmuA and cmuB were essential for growth of *M. chloromethanicum* CM4 on MeCl. These additional genes, metF (encoding methylene-H_4F reductase), folD (encoding methylene-H_4F dehydrogenase/methenyl-H_4F cyclohydrolase), and purU (encoding formyl-H_4F hydrolase) were specifically induced during growth in the presence of MeCl and were transcribed from consensus promoters that differed from previously identified *Methylobacterium* promoters.

On the basis of these genetic and biochemical data, a new C_1 utilization pathway was postulated for growth of CM4 on MeCl, which in part is similar to the H_4F-linked pathway of formaldehyde oxidation. In this pathway, MeCl is dehalogenated by CmuA in the initial step and the methyl group is transferred to a corrinoid on the same polypeptide chain. CmuB, the secondary methyltransferase, then transfers the methyl group to H_4F yielding methyl-H_4F, which is further oxidized by 5,10-methylene-H_4F reductase to 5,10-methylene H_4F. This latter

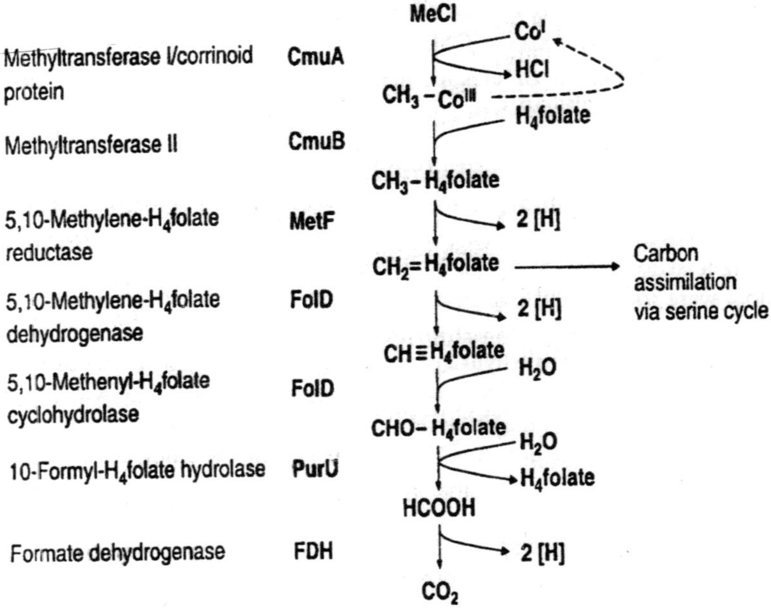

Fig. 14.2. The MeCl utilization pathway of Methylobacterium CM4 as proposed by Vannelli and coworkers.

intermediate provides the formaldehyde for carbon assimilation by the serine cycle which is critical for the synthesis of multicarbon compounds from C_1 substrates. However, 5,10-methylene H_4F can also be further oxidized to yield reducing equivalents for energy generation. This is accomplished by the bifunctional enzyme 5,10-methylene H_4F dehydrogenase/5,10-methenyl-H_4F cyclohydrolase (FolD) yielding 10-formyl-H_4F. In the final steps of the pathway, 10-formyl-H_4F is oxidized to formate by 10-formyl-H_4F hydrolase and formate is oxidized to carbon dioxide by formate dehydrogenase. It is interesting to note that some of the potential energy gain from the oxidation of 10-formyl-H_4F is "wasted" by the hydrolase reaction, since the oxidation of this compound could also be coupled to generation of ATP by the enzyme formate-H_4F ligase (formyl-H_4F synthetase), if present in this organism.

MeCl metabolism in A. lissarensis CC495

A 67-kDa halomethane:bisulfide/halide ion methyltransferase that was induced in *A. lissarensis* CC495 during growth on MeCl was purified by Coulter et al. (1999). The polypeptide transferred methyl groups from MeCl, MeBr, and MeI to acorrinoid bound on the same polypeptide chain. The polypeptide also exhibited transhalogenating activity by transferring the methyl group from methyl halides to a number of acceptor ions including I^-, HS^-, Cl^-, Br^-, NO_2^-, CN^-, and SCN^-. It was suggested that HS could be the physiological acceptor ion for methyl groups from methyl halides, since activities of methanethiol oxidase and formaldehyde dehydrogenase were detected in cell extracts of strain CC495 grown on MeCl, providing a route of C_1 metabolism to formate. Gene cloning and sequencing has identified this halomethane:bisulfide/halide ion methyltransferase as a homologue of CmuA. Further genes implicated in methyl halide metabolism in *M. chloromethanicum*, *H. chloromethanicum*, and *A. ciceronei* were found upstream and downstream of the gene encoding CmuA in strain CC495. These findings suggest that methyl halide metabolism in strain CC495 might also proceed via H_4F, or alternatively, that different pathways for methyl halide degradation may exist even in the same organism which might operate under different physiological conditions. This needs to be addressed in more detail in future studies.

Cloning and Sequencing of cmu Gene Clusters

To date, cmu gene clusters have been cloned and sequenced from six methyl halide-degrading strains, including *M. chloromethanicum* CM4, *H. chloromethanicum* CM2, *A. ciceronei* IMB-1, *A. lissarensis* CC495, "*Roseovarius*" strain 179, and "*Ruegeria*" strain 198. A high degree of conservation of the structure of cmu clusters between methyl halide degrading bacteria has been observed. In most cases, the order of genes found in cmu clusters is conserved to some degree, the exception being *M. chloromethanicum* CM4 where cmuB and cmuC are on one transcriptional unit and cmuA is on a separate transcriptional unit. A different organization of cmu genes was found in *H. chloromethanicum* CM2, *A. ciceronei* IMB-1, *A. lissarensis* CC495, and strains 179 and 198. Apart from *H. chloromethanicum* CM2, in which folD was found upstream of a cmu cluster, the gene clusters did not contain cob genes of cobalamin biosynthesis, folC and folD, nor purU encoding formyl H_4F hydrolase. However, this does not exclude the possibility that these may be located upstream or downstream of the partially sequenced gene clusters from these isolates.

Sequencing of the cmu gene clusters from *H. chloromethanicum* CM2, *A. ciceronei* IMB1, *A. lissarensis* CC495, strains 179, and 198 identified additional genes, fmdB, paaE, and hutI and the order of these genes was conserved. These genes were hypothesized to encode a transcriptional regulator (fmdB), a reductase (paaE), and an imidazolone propionase (hutI). Although the role of these genes and

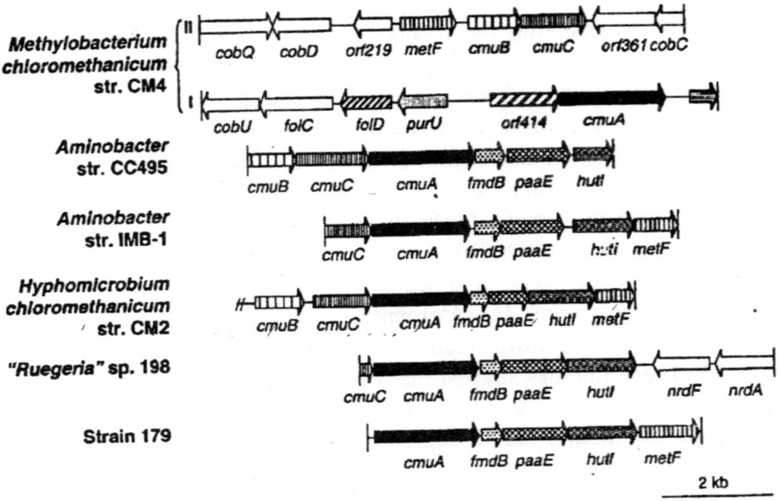

Fig. 14.3. Gene clusters of methyl halide-utilizing strains containing genes with proven and putative function in the CmuA methyl halide utilization pathway.

their encoded polypeptides is unclear, it is possible that they have a role in the metabolism of methyl halides. Of the three additional open reading frames upstream of cmuB that were identified in *H. chloromethanicum* CM2, one was similar to a porin precursor from *Neisseria denitrificans*. The other two, however, did not have matches in the database.

As in *M. chloromethanicum* CM4, a gene (metF) encoding methylene H_4F reductase was found downstream (3′) of hutI in *A. ciceronei* IMB-1, *H. chloromethanicum* CM2, and the marine strain "*Roseovarius*"sp. 179. In the marine strain "*Ruegeria*" sp. 198, metF was not found and sequence downstream (3′) of hutI contained ribonucleoside diphosphate reductase genes nrdF and nrdA orientated in the opposite direction to the open reading frames of the cmu gene cluster.

The location of metF is therefore not conserved between these strains. In four of these gene clusters, cmuC was found upstream of cmuA, and additionally cmuB was found upstream of cmuC in *A. lissarensis* CC495 and *H. chloromethanicum* CM2. The high similarity between these gene clusters suggests that the gene order cmuBCA may be conserved between all these gene clusters with the exception of M. chloromethanicum CM4. Further, cloning and sequencing work is required to investigate this in detail.

Mutational and Transcriptional Analysis of cmu Genes of *H. Chloromethanicum* CM2

Initial cloning and sequencing of the cmu gene cluster of *H. chloromethanicum* CM2 was accomplished by McAnulla et al. (2001b). Further progress was made by Borodina et al. (2004) who successfully developed and applied a genetic system for strain *H. chloromethanicum* CM2, allowing delivery of transposon mutagenesis vectors by electroporation. Borodina and colleagues obtained a number of miniTn5 insertion mutants that could not grow on MeCl. In one of these mutants, the transposon was inserted at the 3′ end of a gene encoding the putative imidazolone propionase HutI and 5′ of the start of the gene encoding methyl H_4F reductase (MetF) which overlapped with the end of hutI. A further eight transposon insertion mutants disrupted the gene encoding the putative methyltransferase CmuC. Although insertional inactivation of cmuC also produced a MeCl-minus mutant in *M. chloromethanicum* CM4, in *H. chloromethanicum* CM2, the mutation affected expression of CmuA as shown by SDS-PAGE analysis of cell extracts of the mutants. This was different to the findings in *M. chloromethanicum* CM4, where CmuA and CmuB were both expressed and functional in cmuC$^-$ mutants. However, mutation of cmuC led to a MeCl-minus phenotype in *M. chloromethanicum* CM4, suggesting that the CmuC methyltransferase has a role in methyl halide metabolism which has yet to be determined. In *H. chloromethanicum* CM2, the mutation in the hutI-metF region had a similar effect on expression of CmuA, again revealed by SDS-PAGE analysis, as did the mutation of cmuB (accomplished by marker exchange mutagenesis). This led to the suggestion that mutations in cmuB, cmuC, and hutI-metF exerted polar effects that abolished the expression of CmuA. Reverse transcriptase PCR experiments showed that the cmuBCA-metF operon was cotranscribed in *H. chloromethanicum* CM2 and tightly coregulated in a strictly MeCl-dependent manner, whereby the presence of MeCl led to the transcription of the cluster, even when methanol was also present during growth. The promoter of the operon has subsequently been identified by RACE PCR (rapid amplification of cDNA ends) and lies upstream of cmuB.

Evidence for Operation of the Cmu Pathway in MeCl$^-$ and MeBr-degrading Bacterial Isolates

There is ample evidence that the isolates harboring cmu gene clusters degrade methyl halide using a pathway similar to that identified in *M. chloromethanicum* CM4. Mutational analysis of cmu cluster genes (i.e., cmuB, cmuC, cmuB, metF, and purU) in CM4 and *H. chloromethanicum* CM2 has shown that these genes are required for growth on methyl halides. The enzyme activities of this pathway have been demonstrated in CM4 and CM2 and CmuA has been purified from *M. chloromethanicum* CM4 and *A. lissarensis* CC495. Expression of CmuA during growth on methyl halides has been demonstrated by SDS-PAGE in the soil isolates *M. chloromethanicum* CM4, *H. chloromethanicum* CM2, and *A. lissarensis* CC495. CmuA was identified as a 67-kDa polypeptide that was expressed during growth on MeBr in the marine isolates "*Roseovarius*" strain 179 and "*Ruegeria*" strain 198. This has been confirmed by mass spectrometry.

Identical polypeptide profiles of biomass grown on MeCl and MeBr of "*Ruegeria*" strain 198 confirmed that the same pathway is responsible for degradation of both MeBr and MeCl. In *H. chloromethanicum* CM2, transcription of the cmu gene cluster was controlled by MeBr in a similar fashion as with MeCl. Although no detailed investigation of MeI degradation has been reported for any of the methyl halide-degrading isolates, it seems reasonable to assume that the CmuA pathway could also be responsible for degradation of MeI, especially since it has been shown that CmuA from *A. lissarensis* CC495 was able to transfer the methyl group from MeCl, MeBr, and MeI to a variety of acceptor ions.

It is likely that environmental concentrations of methyl halides are sufficient to induce the enzyme system, although it might be difficult to observe the 67-kDa band of CmuA by SDS-PAGE when cells are grown at atmospheric methyl halide-mixing ratios, since the amount of this peptide may be low under such growth conditions. Whether natural picomolar concentrations of methyl halides are indeed sufficient to induce the CmuA pathway could be determined by comparison of wild-type and mutant strains of methyl halide utilizers. At present, mutants are only available for *M. chloromethanicum* CM4 and *H. chloromethanicum* CM2. Mutational analysis has not yet been accomplished for the *Aminobacter* strains.

It is important to note that growth of methyl halide-degrading bacteria on more favorable carbon sources did not suppress methyl halide oxidation. *H. chloromethanicum* CM2 could grow on methanol and MeCl simultaneously and expression of CmuA was shown by SDS-PAGE. Similarly, the expression of CmuA in the marine strain 179 growing on betaine and MeBr was demonstrated by mass spectrometry.

Alternative Methyl Halide Degradation Pathways

There is some evidence to suggest that the CmuA pathway is not the only methyl halide utilization pathway. Extensive Southern blotting and gene probing of *L. methylohalidivorans* MB2 with cmuA probes, prepared from terrestrial isolates and from a closely related marine isolate, "*Roseovarius*" 179, failed to show any indications of a homologue of cmuA in MB2; it is therefore possible that *L. methylohalidivorans* MB2 has a different pathway for methyl halide degradation. Another *Roseobacter* clade isolate, "*Roseovarius*" 217 strain did not induce a 67-kDa polypeptide during growth on methyl halides. As with *L. methylohalidivorans* MB2, Southern blotting and gene probing failed to highlight a cmuA homologue in strain 217. Strain LIS-3 isolated by Hoeft et al. (2000) appeared to degrade MeBr by a pathway involving methyltransferase reactions, this was concluded from the inhibition of MeBr oxidation by chloroform. Coulter et al. (1999) reported that the homologue of the CmuA methyltransferase in *A. lissarensis* CC495 was not inhibited by chloroform. It would therefore seem possible that strain LIS-3 used an enzyme system to degrade MeBr which is different to that in *A. lissarensis*.

Microbial Ecology of Methyl Halide-degrading Bacteria

Development of Functional Gene Markers for CmuA Pathway Methylotrophs

PCR primers for the amplification of cmuA were first developed by McAnulla et al. (2001b) based on cmuA sequences from *M. chloromethanicum* CM4 and *H. chloromethanicum* CM2. The gene cmuA was chosen as a key functional molecular marker for targeting methyl halide-utilizing bacteria since it encodes the enzyme carrying out the first defining step in the methyl halide utilization pathway. The unique domain structure of CmuA was exploited by generating PCR primer sequences that would anneal to target sites in regions coding the N-terminal methyltransferase domain and the C-terminal corrinoid-binding domain. This was predicted to increase the specificity of the PCR primer set. While specific PCR products were obtained with six methyl halide-degrading *Hyphomicrobium* isolates, the primers failed to amplify a product from *A. ciceronei* IMB-1. From Southern blotting and gene probing this strain was known to possess cmuA but its sequence was not yet known. New PCR primers were designed, again placing forward primers in the methyltransferase domain and reverse primers in the part of cmuA encoding the corrinoid-binding domain. Sequencing of PCR products confirmed that one of these primer sets amplified cmuA fragments from all *Hyphomicrobium* strains and *A. ciceronei* IMB-1. No amplification was observed with non-methyl halide-degrading isolates. This primer pair was used to amplify cmuA genes from an MeCl-degrading soil enrichment culture. Cloning and sequencing revealed that three sequence types were obtained, one being closely related to *H. chloromethanicum* CM2, another identical to the sequence from *A. ciceronei* IMB-1, and a third sequence that was only 91% identical to CmuA from *H. chloromethanicum* CM2.

PCR primers for cmuA were further refined with the availability of the cmu gene clusters from *A. ciceronei* IMB-1 and *A. lissarensis* CC495 which were also cloned and sequenced and have since proven useful in the analysis of the diversity of this gene in various enrichment cultures, bacterial isolates, and environmental samples. The phylogenetic tree illustrates the diversity of the gene encoding CmuA from various bacterial isolates and sequences obtained after direct amplification of cmuA from a variety of environmental samples.

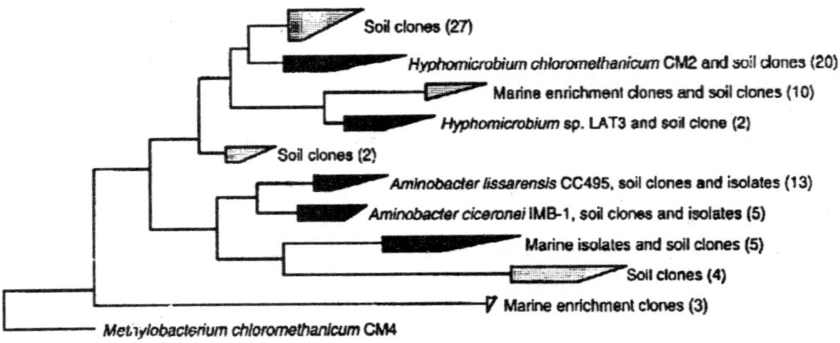

Fig. 14.4. Phylogenetic tree based on neighbor joining analysis of amino acid sequence alignment of CmuA sequences from methyl halide-degrading bacteria and environmental samples.

Stable Isotope Probing of Methyl Halide Degradation in Soils

Stable isotope probing (SIP) is an experimental method that links the degradation of ^{13}C-labeled growth substrates to the phylogenetic diversity of bacteria that use these substrates in situ. Previously, Boschker et al. (1998) used stable isotopes to label phospholipid fatty acids as biomarkers of microorganisms in anoxic sediments. SIP targeting nucleic acids was first developed as DNA-SIP (stable isotope probing based on analysis of DNA) with methanol-utilizing methylotrophs. Since then, the application of SIP has been further developed using both DNA and RNA and a variety of ^{13}C-labeled substrates. The principle of SIP is based on the assimilation of the "heavy" carbon into the biomass of the active population in the environment, which is supplied as >99% ^{13}C-labeled substrate in microcosm experiments. The biomass of ^{13}C-substrate assimilating organisms becomes ^{13}C-labeled, including the nucleic acids; organisms that do not utilize the "heavy" substrate remain unlabeled. The nucleic acids of labeled and unlabeled organisms can be physically separated by density gradient centrifugation and "light" and "heavy" DNA or RNA fractions can then be studied using a variety of molecular microbial ecology methods.

Stable isotope probing of agricultural soils

DNA-SIP was used successfully with MeCl and MeBr in agricultural soils to identify active populations of methyl halide-utilizing bacteria. Light and heavy density gradient fractions of DNA extracted from soil microcosms that had been exposed to ^{13}C-MeCl or ^{13}C-MeBr were analyzed for bacterial diversity using denaturing gradient gel electrophoresis (DGGE) of partial bacterial 16S rRNA genes and by sequencing of cloned 16S rRNA genes. The functional diversity of methyl halide-utilizing organisms was analyzed by clone library analysis of cmuA genes. DGGE analysis of heavy DNA fractions revealed different community fingerprints for microcosms that had been exposed to MeCl and MeBr, indicating that the populations degrading these substrates in situ may be different. Although this might not necessarily be expected, based on the observation that the CmuA pathway is capable of MeBr and MeCl degradation, the fact that not all methyl halide-degrading bacteria are equally capable of growth on MeCl and MeBr (and MeI) suggests that bacteria may have differing capacity to deal with these substrates of differing toxicity. Analysis of 16S rRNA gene clone libraries of the heavy DNA fraction failed to show sequences matching those of known methyl halide-utilizing isolates, such as *Methylobacterium*, *Hyphomicrobium*, and *Aminobacter*. The dominant sequence types that were recovered from the heavy DNA fractions of MeBr soil microcosms were related to *Burkholderia*, while the dominant clone sequences in heavy DNA fractions of MeCl soil microcosms were related to *Rhodobacter*, *Lysobacter*, and *Nocardioides*. Further phylotypes were detected in both MeCl and MeBr microcosms; however, the diversity detected by the clone libraries was greater for MeCl soil microcosms compared to MeBr microcosms, possibly due to the higher toxicity of MeBr. The detection of 16S rRNA gene sequences similar to *Nocardioides* was particularly interesting since the sequence was similar to a methyl halide-degrading isolate previously isolated from woodland soil which was lost, confirming that these bacteria may play a role in cycling of methyl halides. The results of the study by Miller et al. (2004) demonstrated that the dominant types of methyl halide-utilizing bacteria in agricultural soils may be different from the extant isolates, and suggested that the phylogenetic diversity of methyl halide-degrading bacteria is not represented by the strains currently in culture. The diversity of cmuA genes in the heavy DNA fraction was also analyzed. The cmuA sequences obtained from the SIP experiment belonged to four distinct clades, which were similar to cmuA from *A. lissarensis* CC495, a second group of sequences more distantly related to cmuA sequences from *Aminobacter* species, a deeply branching group of sequences, and one clone which was similar to cmuA sequences of *Hyphomicrobium* species. The observation that the overall diversity of 16S rRNA libraries was higher than that of the cmuA clone libraries might be related to a certain bias in the primer sequences that were designed based on the known cmuA genes from isolates that only represented *Alpha-Proteobacteria*. They might therefore fail to amplify more divergent cmuA genes. Furthermore, it is possible that some of the methyl halide-utilizing populations, detected in 16S rRNA gene libraries, could have a different pathway of methyl halide degradation.

Stable isotope probing of woodland and forest soils

In a similar study, Borodina and colleagues used DNA-SIP to analyze the diversity of MeCl-utilizing populations in a variety of soils. The SIP approach was complemented by enrichment and isolation of MeCl-degrading bacteria. Isolates were identified by sequencing of their 16S rRNA genes and included strains closely related to *Aminobacter* and *Mesorhizobium*. The majority of the isolates however were identified as *Hyphomicrobium* species. The diversity of cmuA was also analyzed using DNA directly extracted from soil samples, DNA extracted from enrichment cultures and ^{13}C-labeled DNA from SIP experiments. A major group of cmuA sequences belonged to a clade that did not have cultivated representatives, and included cloned cmuA sequences obtained from DNA of woodland, forest, and garden soil, as well as cmuA clones from enrichments inoculated with garden soil. Sequences of cmuA closely related to cmuA of *Aminobacter* were identified in DNA from garden soil enrichments and DNA extracted from wood soil. Another clade was composed of *Hyphomicrobium* species cmuA sequences and related sequences which were detected in garden soil DNA and also as a dominant cmuA sequence type in clone libraries of the heavy fraction of a wood soil DNA-SIP experiment. Surprisingly, sequences of other dominant cloned cmuA genes from the SIP heavy fraction were similar to those of marine cmuA sequences obtained from MeBr enrichment cultures of coastal seawater from the English Channel. Further cmuA sequences from soil identified by SIP were similar to that of a marine isolate, strain "*Roseovarius*" 179, and somewhat more distantly related to cmuA sequences from the SIP experiments with soil carried out by Miller et al. (2004).

The results of this study suggested that *Hyphomicrobium* spp. may be active in particular soils, but also that they are dominant isolates because they are favored by the enrichment and isolation conditions used in these experiments. Other dominant types of MeCl-utilizing bacteria (indicated by direct cloning and sequencing and by SIP experiments) have yet to be cultivated from soils and need to be identified and characterized at the physiological level. The high similarity of cmuA sequences from marine enrichments and isolates and certain cmuA genes detected by DNA-SIP in soils was unexpected and an interesting observation.

Marine Methyl Halide Degradation

Relatively little is known about the diversity of methyl halide-degrading bacteria in seawater. Until now, the identification of methyl halide-degrading bacteria has mainly been restricted to isolation of strains that could grow on MeBr or that were able to cometabolize it. *L. methylohalidivorans* MB2, "*Roseovarius*" sp. 179, "*Roseovarius*" sp. 217, and "*Ruegeria*" sp. 198 were all affiliated with the *Roseobacter* clade, but methyl halide utilization is not a general trait of *Roseobacter* group isolates. Analysis of the diversity of cmuA genes obtained by PCR amplification, cloning, and sequencing directly from community DNA extracted from seawater samples and methyl halide-degrading enrichments indicates that additional uncultivated lineages of cmuA-containing organisms are present in seawater. A report by Goodwin et al. (2005) indicates that MeBr may be degraded in the marine environment by bacteria, such as *Sphingomonas* Oxy6, which co-oxidized MeBr using an enzyme that was inhibited by toluene.

POTENTIAL APPLICATIONS FOR BIOREMEDIATION USING METHYL HALIDE-OXIDIZING BACTERIA

Reducing Fugitive MeBr Emissions

Most of the anthropogenic use of MeBr (~ 50 ktonnes per year) is for biocidal purposes. Of this amount, the majority (70%) is used in shallow fumigation of agricultural soil for control of nematodes, fungi, or weeds prior to planting high-value crops such as tomatoes or strawberries. An additional 25% is used to fumigate commodities, such as nuts or grains, during storage or to eradicate pests during quarantine associated with worldwide shipping of produce or wood products. Only 3% is used for structural fumigation for pest control. Not all of the MeBr applied during these commercial activities is released to the atmosphere, however much of what is released may be controlled by the application of various physical, chemical, or biological barriers. Here we examine the potential use of methylotrophic bacteria, which are emphasized throughout this review, to decrease or eliminate the emission of anthropogenic MeBr to the atmosphere.

Previous Efforts to Reduce MeBr Emissions

Before this discussion, it is worth pointing out that other strategies for ameliorating the flux of MeBr to the atmosphere have been proposed and tested with varying results with regard to their overall efficacy and economic practicality. It has long been known that application protocols affect the emission of MeBr during agricultural field fumigations. As much as 90% of the added MeBr may escape to the atmosphere under certain conditions. Injection of fumigant deep in the soil (>30 cm) and optimization of soil moisture content can reduce emissions, while emplacement of virtually impenetrable film (VIF) over the field during the fumigation period can nearly eliminate losses during the covered period. Compounds that react rapidly with MeBr (e.g., ammonium thiosulfate) may provide an effective chemical barrier when applied to the soil surface prior to fumigation. However, of these mitigation strategies, only deeper injection and premoistening the soil have been adopted as standard practice. Either costs or environmental concerns have, so far, limited the adoption of VIF tarps or chemical flux barriers.

Bioremediation using Methyl Halide-oxidizing Bacteria

Direct application of methylotrophic bacteria to soil

Use of methyl halide-oxidizing methylotrophs to enhance the natural degradation of fumigant MeBr in soils was first proposed by Connell Hancock et al. (1998). They observed accelerated degradation of MeBr in soil microcosms inoculated with *A. ciceronei* (then called strain IMB-1). They further showed that IMB-1 could grow on a variety of substrates, including other methyl halides, methyl amines, acetate, and glucose. However, the ability of IMB-1 to oxidize MeBr was constitutive in methyl halide-grown cells but required induction in cells grown on other substrates. Growth on methyl halides was slow and because of the inherent toxicity of these gases, growth could be sustained only if these substrates were added incrementally as discrete pulses to avoid the toxic effects of high single doses. To circumvent this constraint, it proved useful to first grow cells to high density using either glucose or methyl amines and later induce methyl halide oxidation enzymes by exposing them to MeBr.

Practicality of applying bacterial cells directly to surface soils in order to oxidize MeBr was examined during field fumigation of strawberry fields in Monterey, CA. Cells of *A. ciceronei* suspended in minimal phosphate media (3×10^{12} cells m^{-2}) were applied to the soil surface using a garden sprayer immediately prior to fumigation. A mixture of MeBr and the warning agent and biocide chloropicrin (CCl_3NO_2), in a ratio of 67% to 33% w/w, was injected by a tractor moving along the surface. A 0.025 mm thick polyethylene sheet was simultaneously rolled over the surface and covered the field. MeBr concentrations in the soil were determined during the first 4 days of fumigation at three depths below and one depth above the soil surface beneath the plastic sheet. Following the 6 day fumigation period, the sheets were removed and accumulated Br, the product of both biological and chemical mineralization of MeBr, was extracted from the soil using deionized water. More MeBr was mineralized in soils that

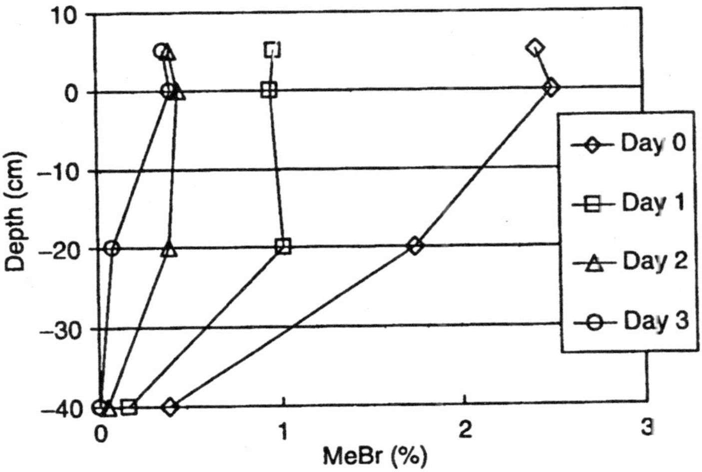

Fig. 14.5. MeBr concentrations in soil gas during the first 4 days of field fumigation with MeBr (67%) and chloropicrin (33%).

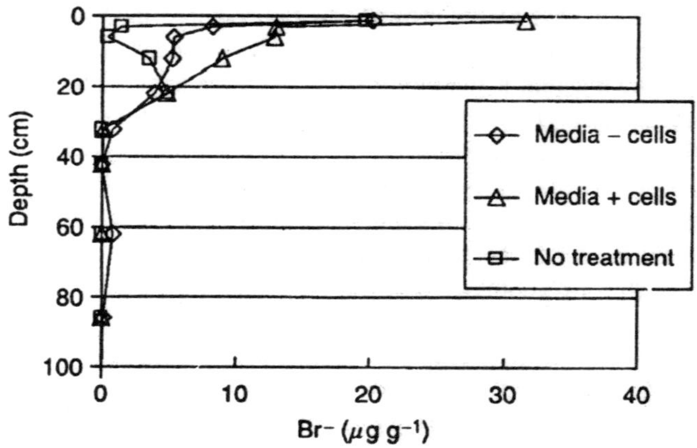

Fig. 14.6. Bromide ion (Br⁻) extracted from soil fumigated with MeBr and chloropicrin (6 days) collected under various soil surface treatments.

received media (with or without cells) than in the controls, indicating that some mineralization was due to nucleophilic reaction of MeBr with water (hydrolysis) or dissolved salts (halide exchange) in the media. A greater amount of MeBr was mineralized in soils that received media plus *A. ciceronei*, and this additional amount was attributed to oxidation by the bacteria.

Enhanced mineralization of MeBr resulting from addition of media was ubiquitous; however, addition of bacterial cells did not always stimulate oxidation. Several factors may have caused an inhibition of the desired bacterial activity, including (1) locally elevated concentrations of MeBr ($>5\%$) beneath the plastic sheet, resulting in direct toxicity to the added A. ciceronei cells, (2) elevated temperatures ($>40°C$) beneath the plastic sheet, exceeding the limits of *A. ciceronei*, and (3) high concentrations of chloropicrin, a known inhibitor of MeBr oxidation by *A. ciceronei*. These three variables were deemed difficult to control during commercial fumigations, hence direct application of methylotrophic bacteria for the purpose of providing a biological barrier to MeBr flux was abandoned, with the recommendation that if and when these variables could be controlled, application of IMB 1 to the soil surface could greatly reduce MeBr emission during field fumigations.

Using bacteria to remove MeBr from the waste stream of contained fumigations

Several nonbiological strategies that use ammonium thiosulfate to scrub MeBr emitted from field fumigations have been proposed.

Chitwood and Deshusses (2001) envisioned collection of MeBr between two layers of plastic sheet during field fumigations and transport by fans to reaction chambers containing a chemical scrubber. Value Recovery, Inc., a New Jersey company, is testing a version of this idea using an additional proprietary catalyst in the scrubber. Other proposed strategies rely on physical trapping of MeBr using activated charcoal or zeolite materials but these methods store MeBr and do not destroy it. They may be useful when chemical or biological destruction can be achieved off site at some convenience.

Some of the above mentioned nonbiological strategies have been suggested or are already in use to remove MeBr from quarantine and pre shipment (QPS) fumigations. Physical traps (e.g., activated charcoal) require further treatment (usually heating) to release MeBr for recycling or destruction. These and the chemical scrubbers used commercially to treat QPS fumigations are subject to breakthrough of MeBr once the number of active sites or the kinetics of the aqueous phase reaction is exceeded. In addition, they have difficulty complying with proposed USEPA regulations limiting maximum emissions to <500 ppm MeBr during operation. Most of these problems can be overcome using bioreactors which utilize the activity of MeBr oxidizing bacteria to completely eliminate emissions from QPS fumigations. Products of laboratory bioreactors using *A. ciceronei* were CO_2 and Br, and the reactors functioned at 100% efficiency for long periods of time when pH and temperature were controlled. Scaling up the laboratory bioreactors in order to demonstrate the economic use of bacteria to destroy MeBr in QPS fumigations presents simple engineering challenges but as of this writing has not yet been attempted.

Conclusions

Genomics of Methyl Halide Degrading Bacteria

The study of methyl halide degradation pathways will further benefit from the genome sequencing projects for *M. chloromethanicum* CM4 and *Roseovarius* sp. 217. In the latter, methyl halide degradation occurs by an as yet uncharacterized pathway. Preliminary analysis of the genome sequence indicates the presence of a number of enzymes involved in C_1 metabolic pathways, for example certain enzymes of the serine cycle of formaldehyde assimilation. The genome sequence does not contain cmuA, thus confirming results of earlier Southern hybridization and PCR analyses. Mass spectrometry of proteins induced during growth on methyl halides is currently being carried out to identify candidate genes that could play a role in metabolism of methyl halides.

Contribution of Alternative Pathways of methyl Halide Degradation

One of the future challenges will be to assess the contribution of the various pathways of methyl halide degradation to the overall biological sink. The work by Goodwin et al. (2005) demonstrates that toluene is a potent inhibitor of MeBr oxidation in the marine environment. It remains to be assessed whether toluene may also be an inhibitor of methyl halide oxidation in extant marine methyl halide degrading isolates other than *L. methylohalidivorans*, which was not inhibited by toluene. Degradation of methyl halides by bacteria is most probably carried out by a number of different pathways. A better understanding of bacterial cycling of methyl halides will require additional research to characterize the unknown methyl halide degradation pathways, for instance in *Roseovarius* sp. 217 and *L. methylohalidivorans*.

While the contribution of individual pathways will only be assessable once suitable inhibitors have been identified, including those that inhibit the CmuA pathway, functional gene markers will need to be developed in order to address the functional diversity of methyl halide utilizing bacteria that do not employ the CmuA pathway of methyl halide degradation. Metabolomics studies using methyl halides labeled with stable or radioactive isotopes coupled to nuclear magnetic resonance analysis might offer a chance to follow the fate of the methyl carbon in order to assess the dominant assimilation pathways for methyl halides.

Evolutionary Aspects and Role of Methyl Halides as Substrates in the Environment

The detection of closely related cmuA sequences in forest soils and marine isolates is intriguing. At least in some of the marine isolates, there is currently no evidence for additional enzymes of primary methylotrophic metabolism since thus far, the methyl halides are the only C_1 substrates that support growth of strains "*Rosovarius*" sp. 179 and "*Ruegeria*" sp. 198, out of the wide range that have been tested. Future genome sequencing of methyl halide utilizing isolates might help to answer the question whether methyl halide degradation pathways first evolved in the terrestrial or the marine environment.

The fact that a number of the known genes involved in methyl halide degradation are evolutionary related to genes encoding enzymes of C_1 metabolism in methanogenic archaea further underlines the possibility that horizontal gene transfer may have played an important role in the diversification of C_1 compound metabolism and the spread of methyl halide utilization genes.

Acquisition and retention of genes for methyl halide utilization probably confers a selective advantage to organisms in environments where methyl halides are produced (e.g., soils, the phycosphere). It has been suggested previously that it is unlikely that methyl halide degrading organisms could live on atmospheric levels of MeBr, but since tropospheric levels of MeBr were readily consumed by *A. ciceronei* IMB-1 and *L. methylohalidivorans* MB2, it is likely that methyl halides can be exploited by facultative methylotrophs as supplementary carbon sources in both soil and marine environments. Bacterial consumption of methyl halides would thus supply extra energy (and could potentially also lead to carbon assimilation) while the cell is using other growth substrates. This does not rule out the possibility of micro sites in the environment that might have higher concentrations of methyl halides that could potentially support growth. Such microsites might exist in soils in proximity of methyl halide-producing fungi and plants, or in the marine environment in the phycosphere of phytoplankton producing MeBr.

15

AMPLIFICATION

A critical component of limiting bacterial disease outbreaks is the tracing of the infection to the index source, which can be facilitated by using a highly discriminating bacterial identification system that will reliably identify genetically related bacterial populations. For pathogenic bacteria with highly conserved genomes, such as the zoonotic pathogen *Brucella*, finding distinguishing markers or traits for strain identification is challenging. This chapter describes a relatively new procedure for identifying *Brucella* strains. The procedure, which is called "HOOF prints" (hypervariable octameic oligonucleotide fingerprints), is based on high levels of polymorphism observed at several genomic loci in the *Brucella* genomes that contain small tandemly repeated deoxyribonucleic sequences. The technique described is designed for medium- to high-throughput analyses. However, the method described can be modified to characterize fewer samples.

HISTORICAL OVERVIEW

Infectious diseases of livestock have a negative impact on the agricultural economy of all countries. A critical component of limiting disease outbreaks is the ability to trace a new outbreak to the original source so that further spread from the index population is prevented. This tracing back requires some type of bacterial identification system that will reliably discriminate between related and unrelated bacterial populations, which is difficult to accomplish for pathogenic bacteria with highly conserved genomes because distinguishing markers or traits are hard to find. *Brucella* species are one example of disease-causing bacteria with highly conserved genomes. Complete genome sequencing of three *Brucella* species has revealed remarkable genetic homogeneity despite differences in host preference. Although some phenotypic characteristics and genetic features have been identified to distinguish among the eight *Brucella* species, highly discriminating methods are limited. Furthermore, some methods have problems with reproducibility from laboratory to laboratory.

Recently, small tandemly repeated DNA sequences have become popular sites for finding genetic polymorphisms in bacterial strains because these short tandemly repeated DNAs mutate at a much higher level than the spontaneous mutation rate, mainly as the result of slip-strand mispairing and recombination. This phenomenon has been known for a long time in eukaryotic organisms and is the basis for human DNA identification, but it has only recently been applied to typing pathogenic bacterial strains, such as *Bacillus anthracis*, *Yersinia pestis*, *Haemophilus influenzae*, and *Francisella tularensis*, to name a few.

This chapter describes in detail the procedure for identifying *Brucella* strains based on the high level of polymorphism found at 10 independent genomic loci that contain small tandemly repeated DNA sequences. Because all of the loci contain 8-bp tandem repeats, we have named this technique "HOOF prints," for hypervariable octameric oligonucleotide fingerprints. The procedure presented here is designed for medium- to high-throughput analyses (at least 10 strains or bacterial colonies) and takes advantage of available technologies for rapid and sensitive genetic analyses, including polymerase chain reaction (PCR) and capillary electrophoresis with fluorescent detection. The methods described can be scaled down to characterize a smaller number of samples and can be performed in a modestly equipped laboratory.

Materials

Laboratory Requirements

1. BL-3 facilities approved for work with *Brucella* (if live *Brucella* will be cultured or used).
2. Dedicated work area completely free of contaminating *Brucella* or *Brucella* DNA.
3. Adjustable pipets dedicated to PCR setup only (e.g., P-2, P-1 0, P-200, and P-1000).
4. Additional pipets (e.g., P-10 and P-200) for dispensing templates and detection procedures.
5. Disposable pipet tips with aerosol-preventing filters.
6. Multichannel pipets.
7. Repeating pipettors.

Sample Preparation

1. Nutrient agar plates such as tryptose agar (Difco), trypticase soy agar (BBL), or *Brucella* agar (BBL) containing 5% serum (bovine, calf, or fetal calf).
2. Saline: 0.85% (w/v) NaCl in sterile water.
3. CO_2 (10%) incubator or jar.
4. Inoculating loops.
5. Spectrophotometer (e.g., Beckman).
6. Disinfectant (e.g., 1% Lysol ICTM, Reckitt & Colman Inc.).
   The following items are needed for methanol preservation of cells.
7. Methanol (reagent grade).
8. Trypticase soy broth with 5% bovine serum.
9. Shaking (rotating) water bath set at 37°C.
10. 50-mL Screw-cap centrifuge tubes.

CR Amplification

1. Thermal cycler with the capacity to use 96-well plates.
2. Tabletop centrifuge with plate rotor (optional).
3. 96-Well PCR plates.
4. Repeating pipettor.
5. Reagents needed for preparing the PCR Master Mixes.
6. PCR-grade water.
7. TE: 10 m*M* Tris-HCl, 1 m*M* EDTA, pH 8.0.
8. 10X Reaction buffer: 500 m*M* Tris-HCl, pH 8.3, 100 m*M* KCl, and 50 m*M* $(NH_4)_2SO_4$ at 25°C.
9. 25 m*M* $MgCl_2$.
10. 2.5 m*M* Each of dATP, dCTP, dGTP, and dTTP (Invitrogen).
11. DNA polymerase.
12. Oligonucleotide primers.
13. PCR cap strips or cap mats.
14. Adhesive plate sealers: foil.
15. Plate roller for sealing adhesive plate sealers, cap strips, and cap mats (optional).

cessing of Amplicons

Agarose electrophoresis system.

BE: 89.0 m*M* Tris-borate, 2 m*M* EDTA, pH 8.3, or TAE: 40 m*M* Tris-acetate, 1 m*M* Na_2EDTA, pH 8.3, electrophoresis running uffer (not required for buffer-less gel systems).

3. Agarose gels, 4%.
4. Multichannel pipettors and compatible filter tips.
5. Ethidium bromide staining solution: 500 μg/L in running buffer or SYBR Gold Nucleic Acid Gel Stain.
6. Loading buffer with tracking dyes.
7. Molecular weight markers: 25-bp DNA ladder and 10-bp DNA ladder.
8. Camera with Wratten no. 22A filter or digital reader for recording gel images.
9. 96-Well PCR plates for diluting amplicons.

Sizing of Amplicons by Capillary Electrophoresis

1. Capillary electrophoresis instrument with fluorescent fragment detector, suitable for small DNA fragment analysis.
2. Deionized formamide.
3. CE running buffer or 10X Genetic Analyzer Buffer with EDTA.
4. Fluorescent internal sizing ladder range 50 to 350 bp.
5. Capillary matrix polymer for fragment separation.
6. Data capture software.
7. 36-cm Capillary array (16 capillaries/array).
8. Computer for data analysis.

METHODS

PCR Setup

The extreme sensitivity of PCR makes this technique highly vulnerable to contamination artifacts. Because a single copy of template can be amplified into significant yields of corresponding products, significant care and planning must be implemented before performing any PCR assays for diagnostic purposes. Preparation of bulk quantities of the reaction mixture should be done in a dedicated room, biological safety cabinet, or enclosed work station that is free of bacteria or DNA.

It is not sufficient to simply wipe down the work area areas with alcohol or certain bactericidal solutions because the DNA of the dead bacteria can still be amplified by PCR. Work surfaces and equipment must be treated with a decontaminant shown to destroy DNA. For example, treatment of surfaces and equipment with 10% bleach will destroy the DNA as well as kill the bacteria as long as all surfaces are exposed. Ultraviolet (UV) radiation also will destroy DNA and is a convenient way to decontaminate biological cabinets or enclosed workstations. Pipets, tube racks, and stock solutions should be dedicated to PCR. Addition of the sample template, amplification, and detection should be performed in another area with a separate set of equipment and supplies than those used for making the master mix. The use of disposable pipet tips containing aerosol-preventing filters is highly recommended for all stages of the procedure.

Sample Preparation

The DNA used for strain typing comes from the bacteria cultured from infected tissues or exudates as part of the normal diagnostic process. All HOOF-print analyses should be performed on individual colonies either directly or subcultured for archiving. Because of the potential for microevolution within the bacterial population, it is highly recommended that several colonies are individually tested for each strain.

1. Cultivate *B. abortus* from abortion material, lymph nodes, milk, and reproductive organs on a primary isolation plate (a basic nutrient-agar plate, such as tryptose agar, trypticase-soy agar, or Brucella agar, containing 5% serum and antibiotics). Incubate in a 10% CO_2 atmosphere at 37°C for 24–72 h.
2. If pure colonies of *Brucella* cannot be obtained from the primary isolation plate, reinoculate suspected *Brucella* colonies onto a secondary isolation plate and incubate at 37°C in a 10% CO_2 atmosphere for 24 to 48 h.
3. Culture suspensions may be preserved in methanol for later use and archiving. Using a sterile inoculation loop, transfer bacteria from a single colony to a flask containing 30 mL of trypticase soy broth with 5% bovine serum. Incubate in a shaking 37°C water bath for 24 to 72 h until a heavy growth is achieved. Transfer the bacteria/broth suspension to a screw-top centrifuge tube that has been weighed. Centrifuge the sample at 7000*g* for 15 min. Discard the supernatant into a bactericidal disinfectant (e.g., 1% Lysol IC) and determine the wet-weight of the pellet. Resuspend the pellet in 100 μL of methanol and 50 μL of saline for each mg wet-weight of the pellet. Incubate the sample at 4°C for at least a week, thoroughly mixing by inversion daily, to ensure complete killing of the bacteria. Samples can be conveniently stored at 4°C in 20-mL scintillation-type vials with screw caps for at least 5 yr without significant degradation or loss of DNA.
4. To prepare bacteria directly from agar plates (primary or secondary) for PCR, use a sterile inoculating loop to transfer bacteria from one large colony to a 2-mL tube containing 100 μL of saline.

5. Before use, remix the culture suspension or methanol preserved suspension and dilute an aliquot 1/10 in PCR-grade water. Mix gently but thoroughly. The diluted material should be appropriately discarded after use.

PCR Amplification

Preparation of multiplex master mixes

Each sample is characterized with eight multiplexed PCR primer combinations, in which most HOOF-print loci are tested twice for reproducibility. A total of 10 unknown samples along with 1 positive control sample and 1 negative control sample can be analyzed in a 96-well PCR plate format.

1. Prepare a 1000X stock of synthesized oligonucleotide primers in TE buffer.
2. Prepare each of the eight primer-cocktail working solutions according to the protocol. The diluted working solution can be stored in the dark at 4°C for approx 3 d without deterioration.
3. Prepare eight PCR Master Mixes (one for each primer combination). Add each component sequentially. Mix completely after the addition of GC Rich solution and again after the addition of the polymerase.

PCR setup

1. Following the diagram, aliquot 25 μL of Master Mix into the bottom of each well across one entire row. This can easily be done with a positive displacement repeating pipettor.
2. If it will be more than a few minutes until the sample templates are ready to add, cover the plate with an adhesive plate sealer to prevent evaporation and store the plate at 4°C (up to 24 h).
3. When the template samples are ready, remove the plate from 4°C and examine the wells from beneath the plate to see that all wells are filled and that none of the liquid remains on the sides of the well or condensed onto the coversheet. If some droplets of solution are on the sides or tops of the wells, tap the plate gently on the lab bench or pulse briefly in a centrifuge equipped with a plate rotor to collect all the fluid at the bottom. Always be sure that the wells are tightly sealed before tapping or centrifuging the plate to prevent cross-contamination.
4. Add 3 μL of one diluted bacterial strain to each of the eight wells in a column. Fill the first column with a positive control strain or an allelic ladder and then fill the last column with PCR-grade water as the negative control. Be certain that the sample

is deposited into the solution at the bottom of the well. The same tip can be used for the entire column if the tip is wiped with a tissue after each addition to remove any liquid adhering to the outside of the tip.

5. Cover the plate with cap strips, a cap mat, or a fresh adhesive backed sheet of aluminum foil or plastic PCR film. The use of a plate roller is recommended to be certain that each well is completely sealed.
6. The following parameters are used to amplify the target DNA: 95°C for 5 min followed by 30 cycles of 95°C for 30 s, 53°C for 30 s, 75°C for 30 s, and one cycle of 75°C for 60 min.

Processing of Amplicons: Monitoring Amplification Success

1. Prepare two 48-well (double decked) 4% agarose gels. Remove residual buffer from the wells. Be sure that the outside surface of the gel cassette is clean and free of liquid. For electrophoresis in a system that uses a running buffer.
2. Load each well with 5 μL of amplified sample in a total volume of 20 μL. The easiest way to do this is to use an expandable 12-channel programmable pipettor with the following program: pick up 17 μL (water); pick up 5 μL (sample); dispense 20 μL (into gel well); purge. No loading buffer or tracking dye is needed with a bufferless system.
3. In the remaining marker lanes, add a 10-bp DNA ladder and a 25-bp DNA ladder at 250 to 500 ng per lane. For double-decked gels, be sure to include markers in each row of samples.
4. Place the gel cassette in the electrophoresis apparatus and connect the system to the power supply. Run the gels at 70 V for 20 to 30 min at room temperature.
5. If the gel has not been prestained with ethidium bromide, then the gel should be soaked for 10 min in a solution of 250 μg of ethidium bromide in 500 mL of the running buffer, or for greater sensitivity in SYBR Gold Nucleic Acid Gel Stain. If the background fluorescence is too bright under UV illumination, the gel can be destained in distilled water for 20 min.
6. Record the gel image digitally or on film under UV illumination.

Sizing of Amplicons by Capillary Electrophoresis With Fluorescent Detection

Identification of alleles that differ by 8 bp requires a size analysis method that is capable of very precise measurements. Capillary electrophoresis with fluorescent detection has this capability. The high sensitivity allows minute amounts of sample to be sharply resolved by differential migration through a 36-cm electric field.

Sample preparation

Samples should be diluted in PCR-grade water, typically in the range of 1:4 through 1:200, depending on the amplicon concentration, which can be estimated from the intensity of the DNA staining displayed on the agarose gel. If the product band is very light or not visible, a 1:4 dilution is an appropriate starting point. The diluted samples are then prepared for electrokinetic injection into the capillary array by mixing 1.5 μL of the diluted PCR product with 0.25 μL of the internal size standard and 9.75 μL of deionized formamide.

Controls

In addition to a PCR-positive control strain, size determination by capillary electrophoresis is more accurate with a set of allelic ladders or amplicon standards for each primer cocktail. An allelic ladder can be prepared by pooling a number of individually amplified alleles in equalized amounts that span the typical range of allelic sizes at that locus. Amplicon standards are made by pooling the individually amplified alleles from a few, well-characterized strains, in which the individual alleles have been confirmed by DNA sequencing. In this case, the allelic sizes will not be a regularly spaced ladder, but they are easier to prepare than a true ladder.

Capillary electrophoresis conditions

The following conditions apply to the ABI Prism 3100 Genetic analyzer.

Run information:

- Data collection version: 1.0.1
- Analysis Module: GS500 Analysis.gsp
- Run Module: GeneScan36_POP4DefaultModule

Data collection settings:

- Prerun Voltage: 15 kV
- Injection Voltage: 1 kV
- Run Voltage: 15 kV
- Temperature: 60°C
- Prerun time: 3 min
- Injection Duration: 22 s
- Run Time: 25 min

Gel information:

- Capillary separating polymer gel matrix: POP4
- Capillary length: 36 cm
- Run buffer: A.C.E CE Running Buffer - 1X concentration

Data Analysis

The capillary electrophoresis systems with fluorescent detection all use complementary software packages for data capture and analysis. A detailed review of these software packages is beyond the scope of this chapter. However, some familiarity with the data output will be helpful for the interpretation of results. The following briefly describes the basics for using GeneScan software to view and record data from the ABI Prism 3100 Genetic Analyzer.

Creating a project

1. If the data have been saved as a zip or a tar file to conserve space, the file must be extracted with an appropriate program.
2. Open the GeneScan software program.
3. View the menu options under the *File* tab on the main toolbar; highlight the <*New*> option. A pop-up menu will appear with several icons. Single-click the *Project* icon.
4. The Analysis Window will appear.
5. View the menu options under the *Project* tab on the main toolbar of the Analysis Window. Highlight the <*Add sample files*> option. Browse through the directories by double-clicking the appropriate folder icons along the path to the extracted data files (the file names will have an "fsa" extension). Each file contains data for 1 of the 96 samples that were analyzed. Highlight individual files for loading and press the <Add> button. Alternatively, if all the files will be viewed simply press the <Add All> button. The files to be loaded will appear in the second window at the bottom of the screen. If the files listed in this box are the correct files to be viewed, press the <Finish> button.
6. The list of individual files now appears in the Analysis window.

Viewing and recording the data

1. Double click one of the Sample File cells to open a new window containing the electropherogram and a corresponding data table.
2. The electropherogram consists of a series of peaks along the x-axis. The heights of the peaks (y-axis) are measured in relative fluorescent units (RFUs) that were emitted as the DNA fragments passed by the detector. The color of the peak corresponds to the dye color detected in the sample except that black is substituted for the yellow dye (NED) for better visibility. With the HOOF-print multiplex protocol, either three or four colors will appear in each electropherogram. All samples contain a series of red peaks.

These peaks are from the ROX-500 internal size standard used to calibrate the size calling curve. The other peaks identify the alleles amplified with the primer cocktail used with the sample.

3. The results table lists the specific values for the data shown in the electropherogram. The table provides the following information:
   (a) *Dye/Sample Peak:* This column identifies the dye color, and sequentially numbers all of the fragments that fluoresce above the threshold limit (e.g., 50 RFU) set to distinguish peaks from background noise.
   (b) *Minutes:* This displays the time required for each fragment to migrate through the capillary to the detector.
   (c) *Size:* This is the size of the fragment in base pairs calculated from the size curve generated by the internal standards.
   (d) *Peak height:* This is the maximum number of fluorescent units recorded when the fragment passed by the detector.
   (e) *Peak area:* This is the total calculated fluorescent units emitted by the fragment as it passed by the detector.
   (f) *Data point:* This column identifies the position of the peak along the x-axis of the electropherogram.
4. Another notable feature of the electropherograms is the presence of a large peak near the start (left edge) of the chromatogram. This peak contains an assortment of small-sized DNA fragments that run directly through the capillary, relatively unhindered by the gel matrix. The run-through peak is composed mainly of primers and primer-dimers and should be ignored. The data analysis software can be programmed so that the fragment size-threshold for the data table is larger than the sizes of the run-through fragments. As a result, the run-through peak will not be displayed in the table. It is not necessary to record all the data in the data table. The data can be adequately summarized by recording the fragment size in base pairs and the peak height. The peak height is a useful measure of the relative abundance of the fragment. If the peak height is very small (less than 100 units) the corresponding peak might not be significantly greater than the background fluorescence. If the concentration of the allele fragment is too high (e.g., greater than 7000 RFU), the emitted light can overload the sensors, cause overestimation of background products and contaminants, or cause other problems. Emissions in the range of 1000 to 4000 RFU generally produce unambiguous results. A

convenient table layout for recording data can be formatted with a spreadsheet program. Since the spreadsheet contains data for 228 individual alleles, the spreadsheet template can be set up with color-coded fonts to help simplify the data, visually.

Interpretation of Results

HOOF-print genotypes consist of the specific alleles identified from 10 polymorphic genomic loci. Alleles are named for the calculated number of complete repeat units at each locus because each locus also contains a characteristic partial repeat in addition to the complete repeat units. Using Locus-1 as an example, Allele 2 has 2 and 3/8 repeat units, Allele 3 has 3/8 repeat units; and so on. The sizes of the respective partial repeats for the 10 loci are:

Locus-1	3/8
Locus-2	1/8
Locus-3	6/8
Locus-4	1/8
Locus-5	1/8
Locus-6	3/8
Locus-7	3/8
Locus-8	3/8
Locus-9	1/8 + (–1/8)
Locus-10	(–1/8)

Locus-9 contains two series of 8-bp repeats that are different in sequence, are approx 40 bp apart and appear to mutate independently. When choosing the amplification primers for this locus, two primer sets were tested, and it was discovered that the best performing primers amplified both groups of tandem repeats. One of the tandem repeat groups has one extra base pair of the repeat sequence (+1/8), whereas the other series lacks one base pair of the final repeat unit (–1/8). Therefore, Allele 4 may actually consist of 2 and 1/8 repeat units, plus 1 and 7/8 repeat units. Locus 10 also contains a partial final repeat unit that is one base pair short of complete, so that Allele 2 is actually 1 and 7/8 repeat units.

A conversion table for determining the number of repeats at each locus based on the fragment size is presented. This table lists the *theoretical* molecular size in base-pairs calculated from DNA sequence analysis of multiple alleles for each locus. However, in practice, the observed fragment sizes determined by capillary electrophoresis were found to differ from the predicted molecular sizes by as much as 5 to 6 bp. Fortunately, with proper controls and consistent assay conditions, these variations arise in predictable patterns that permit reproducible allele assignments.

A number of reasons exist for the discrepancies between the theoretical allele sizes and the observed sizes. It has been shown that the mobilities of rhodamine-based dyes (ROX and TAMRA) in POP4 polymer-filled capillaries are not consistent with the mobilities of fluorescein dyes (i.e., 6-FAM, HEX, NED, JOE, and TET). The differences can range as high as 6.5-bp for a 100-bp fragment.

The HOOF-print genotypes are written as a sequential numerical string of ordered alleles calculated for Loci 1 to 10. For example, the HOOF print for the vaccine strain, *B. abortus* S19 is: 5, 4, 4, 2, 2, 2, 8, 2, 13, 6, meaning that Locus-1 has Allele 5; Locus-2 has Allele 4; Locus-3 has Allele 4; and so on. This format makes it very easy to construct a multilocus DNA fingerprint database that could be used by any laboratory in the world. The development of an International *Brucella* Fingerprint Database is currently in progress.

Troubleshooting Potential Problems With the Experimental Data

There are a number of things that can affect data production or data interpretation. The following section describes some of the more common problems encountered when performing the HOOF-print assay. This section also shows that it is essential to carefully examine the data before reaching a final determination. One particular danger is the overreliance on allele-calling software that automatically assigns the data into bins (alleles) based on size. These programs may not be able to distinguish between true and false signals and, therefore, it is important to visually inspect and confirm allele identifications.

Abnormal migration of locus-4 alleles

It was noticed early in the development of the HOOF-print assay that *B. suis* isolates exhibit abnormal migration behavior for Locus-4 alleles. Sequence analysis of selected alleles showed that *B. suis* isolates contain a specific mutation in the conserved flanking DNA sequence that increases the size of each allele by 11-bp. Because of the 11-bp change, the *B. suis* alleles for Locus-4 do not match the theoretical allele sizes for *B. abortus* and *B. melitensis* strains. Alleles listed on the alternative chart are designated with an "A" (e.g. Allele 3A). Further studies have shown that *B. canis*, *B. ovis*, *B. neotomae*, *B. cetaceae*, and *B. pinnipediae* strains follow the *B. suis* allele size predictions.

Double peaks caused by nontemplated nucleotides

DNA polymerases (e.g., Taq DNA polymerase) that lack 5′ to 3′ exonuclease activity will add an extra nontemplated nucleotide (typically dATP) to amplified products. However, the addition of the nontemplated base is not absolute; therefore, amplifications performed with a Taq-type polymerase will consist of two populations of amplicons: those with the nontemplated base and those without the extra base. During analysis, these populations will be resolved into a double peak separated by 1 bp. Often, the double peak is not resolved clearly in the electropherogram but instead is listed in the data table as two peaks.

During PCR, the 60-min, 75°C incubation, inserted after the last amplification cycle, drives the process toward maximal addition of the nontemplated nucleotide. However, the nontemplated addition is influenced by the flanking sequence so that some loci produce mainly the "true" allele product whereas other loci produce mostly the "plus-A" product. Nontemplated nucleotide addition is rarely a problem for identifying HOOF-print alleles because allele sizes differ by 8-bp units. However, this is one factor that causes the observed amplicon size to differ from the theoretically calculated size based sequence data.

Stutter peaks

Stutter peaks are caused by imperfect amplicons containing too many or too few repeat units. This phenomenon is caused by slip-strand mispairing during PCR amplification, whereby the DNA polymerase essentially loses its place when copying a large tract of repeats. These amplicons differ from the original DNA template in multiples of 8 bp. Although all variable number tandem repeat (VNTR) loci will produce stutter products during amplification, the proportion of stutter products decreases as the length of the repeat sequence increases. Thus, loci containing 2-bp repeat units have the highest proportion of stutter products. The size of the repeat string also affects the amount of stutter produced. Long strings have a greater chance of slippage than short strings.

Typically, stutter products are of minor concern for HOOF-print analyses because an 8-bp repeat size is not particularly prone to slippage. However, for loci that contain a large number of repeats, or if the assay is performed under nonideal conditions, stutter peaks can be significant and may even obscure alleles from other loci. The straight arrows point to the stutter peaks and the associated true peak for Locus-7. As shown in the corresponding table, the stutter peaks differ in 8-bp intervals (data indicated with a single heavy circle). The enlarged insert shows the stutter peaks more clearly after subtracting the data for Loci-2 and -8. Notice that the stutter radiates out in both directions from the true peak but that the stutter peaks on the negative side (loss of repeat units) are higher (i.e., more DNA) than the peaks on the positive side (addition of repeat units). Also, note that each stutter peak becomes progressively smaller in height (less DNA), the further they are from the true peak.

The presence of substantial stutter peaks can obscure alleles in a multiplexed mixture. The allele for Locus-8 can easily be overlooked among the stutter peaks from Locus-7. When the peaks from Locus-7 are subtracted from the electropherogram, the Locus-8 allele becomes prominent. To a certain extent, color-coding of the peaks helps in sorting through the peaks. However, if the allele from one locus overlaps with stutter from another locus, the data could be difficult to interpret.

Overlapping alleles

One of the problems associated with multiplex analyses is the potential for alleles to overlap, if both alleles are the same size. To minimize potential problems with interpreting data containing overlapping alleles, the primer cocktails were selected to contain primer pairs in different combinations so that if one combination overlaps, there is a good chance that the other combination will not overlap.

Bleed-through peaks and spikes

Amplicon colors are determined by differences in the peak emission wavelengths of the four fluorescent dyes. 6-FAM emits a blue wavelength, HEX emits a green wavelength, NED emits a yellow wavelength, and ROX emits a red wavelength. During electrophoretic separation, the migrating DNA molecules are illuminated by an Argon laser as they pass by the detector, where they are scanned by four independent sensors, each set to detect light emissions of a specific wavelength matching the optimal wavelength of one of the dyes. Bleed-through peaks, also known as "pull-up" peaks, occur when the signal from one dye (e.g., blue) is picked up by the sensor for a different dye (e.g., green) so that the signal is read by both sensors as independent peaks.

Bleed-throughs can occur because of emission spectra of the four dyes overlap so that although most of the light is emitted at the optimal wavelength, small amounts of light are emitted over a wider range of wavelengths. Usually, this is not a problem because the amount of light emitted at wavelengths matching the other dyes is too small to be detected. But if a sample is overloaded, the amount of light emitted at nonoptimal wavelengths may be significant, triggering the sensor to register a nonexisting peak. A clue that bleed-through may be occurring is the presence of a very large peak of one dye color superimposed directly over a smaller peak of a different dye color. Sometimes, these false signals will register with multiple sensors so that multiple dye peaks are superimposed. Bleed-through usually can be resolved by investigating inconsistent results for a given locus when it is multiplexed with different primer combinations. Another clue is that bleed-through peaks may not conform to the expected size of any allele for that locus.

"Spikes" are very sharp thin false peaks, often "multicolored," that may be confused with data peaks. They occur from a number of causes, including micro-air bubbles, dust, urea crystals or dried polymer flakes passing through the capillary, or from minor voltage fluctuations within the instrument. These artifacts may be noticed if the peak in question does not match an expected allele size or if it is found in only one of the duplicate tests for a primer pair. Because spikes are not reproducible in a sample, these artifacts can be identified by rerunning the sample.

Null alleles and microvariants

Sometimes spontaneous mutations will occur at a VNTR locus. These mutations produce alleles of unexpected sizes that are called microvariant alleles. These alleles can be produced by mutations that occur within the conserved DNA that flanks the tandem repeat regions or by irregular mutations that occur within the repeat areas. If the mutation occurs at a primer annealing site, it can prevent amplification of the locus. These are called "null alleles" because no amplicons are produced. Locus-4 is an example of a microvariant locus because an 11 -bp insertion/deletion in the flanking DNA causes the alleles of *B. abortus* and *B. melitensis* to be different from the alleles of the other *Brucella* species. Microvariations have also been observed within Locus-1 and Locus-6.

Microvariants can make the genotyping data difficult to interpret. If an insertion/deletion happens to be a multiple of 8-bp, the locus will be assigned the wrong allele because the mutant product will mimic another allele in size. At the same time, some microvariants can be stable markers that help to identify genetic linkages among strains. For example, an unanticipated allele size at Locus-4 might indicate that the strain is a different species of *Brucella* than expected.

Inaccurate size calling

It is important to choose an internal size standard that covers the entire range of expected allele sizes, since size calling is only accurate for mobilities falling within the calibration curve. A sample calibration curve is shown. The relationship between size and mobility is not linear at the ends of the curve. Therefore, mobilities outside of or near the ends of the calibration curve are not called accurately. There is a gap of data points between approx 200 and 300 bp. This is because the 250-bp fragment runs anomalously in the POP4 separation matrix and is usually omitted from the standard curve calibration. However, without the 250-bp marker, the calibration curve is linear and should be able to adequately determine the size of fragments in this region.

Other causes of migration inconsistencies that can lead to inaccurate size calling involve small differences in the assay environment or machine performance such as run temperature, electrolyte concentration of the sample, and electric field strength as well as the age and condition of the capillaries. Mobility also can be affected by the specific sequence, probably because of the formation of secondary structure, even in denaturing conditions.

Degraded DNA

Degradation of the DNA sample can negatively affect the success of the assay. Some degradation can be tolerated because the amplified products are relatively small. However, significant degradation of the target genomic DNA can cause a PCR amplification bias toward the smallest alleles, whereas larger alleles may not be amplified at all. One clue that indicates potential sample degradation is a noticeable decrease in peak height with increasing allele size at all loci. To prevent degradation of the DNA, samples should be stored at 4°C with a pH of 8.0 to 8.5. If the samples will be stored for the long term, or if it will be more than a few days before the cultured bacteria can be tested, it is best to keep the samples in a preservative that stabilizes DNA.

Identifying microevolution vs superinfection

During the development of the HOOF-print assay, we sometimes found that the capillary electrophoresis data for a particular strain consistently showed multiple peaks/amplicons at one or more loci. Further investigation showed that this resulted from mixed alleles in the sample and was not due to stutter or any other assay-related artifact. There are two categories of mixed alleles detected by this assay. One category is the variant alleles produced by microevolution. The accelerated mutation rates of VNTR loci cause some loci to continuously mutate (adding or losing repeat units). Unless the mutation event is harmful, the strain evolves into a cumulative population of genetically related bacteria with small genotypic differences. Characteristic of microevolution is that the population consists of one predominant genotype with a collection of variant genotypes that differ from the major genotype by a single (or small number of) repeat unit(s). Microevolution can be found within a single infected herd and even an individual animal. The longer the herd is infected, the more microevolution will progress in the bacterial population.

Mixed alleles also can arise from a completely different mechanism, whereby the disease is caused by a mixture of two unrelated strains of *Brucella*. Typically this is recognized by a combination of two very different genotypes in the herd or animal, which at some point in time arose from superinfection with two field strains. In some instances, the mixed strains are different species of *Brucella*. Occasionally, animals are co-infected with a vaccine strain as well as a field strain. These types of mixed infections are most commonly associated with enzootic regions where the disease is prevalent, widespread, and multiple strains persist.

The only way to determine the complete genotypic composition of a strain is to culture the *Brucella* and test multiple colonies, individually. The more colonies that are examined, the more information about strain variants and their frequencies can be determined. Testing the bacteria from multiple positive animals also helps to characterize an outbreak strain.

Notes

1. Separate sets of reusable pipettors must be used for pre-PCR setup and for downstream processes that involve the DNA targets or amplicons to prevent potential cross-contamination of samples or reagents.
2. The use of multichannel and repeating pipettors is not required for this protocol, but they are recommended because they can significantly decrease the amount of time needed to perform the procedure because many pipetting steps are involved and because they decrease the risk of misloading samples into the wrong wells in plates or gels.
3. Preserving samples in methanol is recommended for samples that will not be tested within 7 to 10 d. When preserved as described, samples can be stored at 4°C for at least 5 yr without noticeable

deterioration. As an additional benefit, the bacteria are killed during preservation, reducing the risk of accidental infection of laboratory personnel and eliminating further need for BL3 containment.

4. These plates are sold as fully skirted, half-skirted, and nonskirted, referring to the size of the vertical edge surrounding the outside of the plate. The choice of skirt depends on the brand of thermal cycler and the brand of capillary electrophoresis system used since some brands do not provide enough clearance for all sizes of skirts.
5. When preparing Tris-derived buffers containing EDTA, the Tris component should be dissolved first and brought to a pH of 8.0 before adding the EDTA salt because EDTA is more soluble at a pH value of 8.0 or greater. EDTA is an acid; consequently, the pH value of the solution will need to be checked while the EDTA dissolves and maintained at 8.0. The inclusion of EDTA in certain buffers prevents DNA degradation by chelating Mg^{2+} ions that are essential cofactors for many enzymes, including nucleases. Mg^{2+} ions are also essential for DNA polymerase activity including Taq and other polymerases used for PCR. Therefore, it is important that dilutions of primers and target DNAs are made in water and not TE, since even small changes in the $MgCl_2$ concentration will affect the quality of the amplification reaction.
6. The choice of DNA polymerase is critical to the success of the assay because not all polymerases behave the same. First, the polymerase system chosen should have, or be compatible with, "hot start" capability. That is, the polymerase should be inactivated until the reaction mix is heated to at least 50°C. Second, the polymerase must amplify all targets equally well. A large number of polymerases were tested during the development of the assay with mixed results. Using the parameters specified in this chapter, the best-performing DNA polymerase was FastStart Taq DNA Polymerase by Roche Applied Science; however, other products may be equally suitable.
7. The forward primer for each VNTR locus is labeled with a fluorescent dye at the 5' end. The three dyes (HEX, NED, and FAM) were chosen to be compatible with fluorescent detection by the ABI Prism 3100 Genetic Analyzer with filter set D. If a different brand of fragment analyzer is chosen then the dye labels may need to be changed to meet the requirements of that analyzer.

8. Synthetic oligonucleotide primers should be resuspended for storage in TE buffer at a concentration of 200 pmoles/μL. The primers labeled with a fluorescent dye must be stored in the dark and protected from light as much as possible. Dispensing the primers into 50-μL aliquots will minimize loss if the stock becomes contaminated or mishandled. Synthetic primers, including the fluorescently labeled primers, are stable at 4°C for at least 3 yr if stored in TE and protected from light.
9. Fluorescent-tagged oligonucleotide primers are significantly more expensive than conventional primers. However, the cost is not proportional on the quantity of oligonucleotide synthesized. Therefore, it is only about twice as expensive to order an 80,000 pmol synthesis as it is to order a 10,000 pmol synthesis. Because the primers are stable for at least 3 yr if stored properly, it is more cost efficient to buy large quantities if possible.
10. Plastic cap strips are the most reliable method for sealing the wells of PCR plates, when the thermalcycler does not have a compression style hot bonnet. However, they can be difficult to align and properly seat into the wells. Examine the plate carefully to be sure that all of the caps are seated correctly and completely. It is recommended that the cap-strips be purchased from the same manufacturer as the PCR plates because not all brands of caps will fit into all brands of plates.
11. Be certain that the cap mat or sheet used to seal the plate is designed for PCR and also is designed for the style of plate being used. Some types of plastic film sheets do not seal well on plates with raised edges around the wells. Others cannot withstand the build up of air pressure when the sample is heated, and may partially detach from the wells allowing evaporation of the sample. We recommend using silicone cap mats (DOT Scientific) because they mold to the shape of any well style forming an inverted dome. However, it is good practice to seal the mat with a plate roller immediately before lowering the thermalcycler' s bonnet because the silicone beads sometimes slip out of the wells when not under compression.
12. Many options for gel electrophoresis systems are available. We recommend the use of a bufferless system with precast, prestained gels in plastic cassettes (e.g., E-Gel 48 gels with 4% agarose) because they perform very reproducibly, the plastic cassettes are easy to load with a multichannel pipettor, and no loading buffer

with dye (that can potentially obscure the visibility of bands) is needed for the samples. These systems reduce exposure to hazardous chemicals and also reduce the amount of hazardous waste. However, any standard agarose gel electrophoresis system can be used. When preparing agarose gels, use only DNA grade agarose.

13. SYBR Gold Nucleic Acid Gel Stain is a dye with up to 10 times greater sensitivity than ethidium bromide. However, it cannot be used as a pre-electrophoresis stain because it alters DNA migration and it is absorbed onto glass and certain plastic surfaces. Carefully follow the manufacturer's instructions.
14. Ethidium bromide and SYBR Gold are light sensitive and should be protected from light. Either staining solution (at 1X concentration) can be stored for up to two weeks if stored in a light-proof container (e.g., covered with aluminum foil). Ethidium bromide can be stored at room temperature, whereas SYBR Gold should be stored at 4°C.
15. The quality of the formamide is critical for proper injection and running of the sample through the capillaries. Use only high purity (>99.5%) and low conductivity (<100 μmhos) reagent that has been deionized and stored at an alkaline pH.
16. Most *Brucella* species are Class III pathogens and may fall under special governmental rules and policies. Regardless of the applicable governmental regulations, all steps involving the use of live *Brucella* should be done in an approved Biological Safety Cabinet with appropriate precautions.
17. Better results are obtained if samples containing methanol preserved bacteria are diluted with water and stored at 4°C overnight. However, once it has been diluted in water, the DNA inside the bacteria is less stable and so the diluted samples should not be stored for longer than approx 1 wk.
18. Two loci, Locus-6 and Locus-8, are only tested once per sample. So far, these loci have shown very little polymorphism for *B. abortus* strains. For other *Brucella* species, these loci should also be tested in duplicate.
19. The lyophilized primer should be dissolved in TE buffer as a 1 000X solution. Depending on the information reported in the certificate of analysis provided with custom synthesized primers, (1) add TE to the primer at the ratio, if the certificate reports the total number of micrograms synthesized or (2) add TE at a

ratio of 200 pmol/μL if the certificate reports the total number of picomoles synthesized.

20. The 1 00-μL primer cocktail working solution is sufficient for up to three 96-well PCR plates (30 test samples).
21. VNTR Loci-1 to -8 have a conserved flanking sequence on one side of the repeat sequence, therefore the reverse primers Rev-1 and Rev-3 can be used with multiple forward primers.
22. Thorough mixing of the master mix is critical to the success of the assay because several components are very dense and will settle to the bottom by gravity if not properly dispersed.
23. It is helpful to use a strip of lab tape or masking tape to cover the wells on either side of the well-column to be filled. The tape serves the dual purpose of maintaining orientation and preventing potential contamination of adjacent wells with microdroplets of sample.
24. This protocol uses a higher than normal extension temperature based on empirical optimization of the assay conditions. It is possible that the higher temperature relaxes secondary structure formation in the target DNA, allowing the polymerase to extend through the potentially folded regions.
25. The first incubation step is for activating DNA polymerases that have been chemically modified for hot start capability. The temperature and time should be adjusted for the specific DNA polymerase system used as recommended by the manufacturer. Many DNA polymerases do not need the precycle heating step for activation.
26. Incubating the amplicons for an hour increases the proportion of molecules bearing the nontemplated nucleotide addition by Taq and certain other DNA polymerases and consequently simplifies the fragment separation profile by minimizing one of the peaks in the doublet.
27. Place the agarose gel on the deck of the apparatus before loading the samples. Add 1X running buffer to approx 3 to 5 mm above the gel surface. Load the samples in a high density loading buffer at 1X concentration. If using a multichannel pipettor, be particularly careful not to pierce the gel or the bottom of the wells with the tips, as this can cause the sample to leak out of the well.
28. This program is designed to pick-up 22 μL but only dispense 20 μL to avoid producing bubbles in the wells. However, because the

water is picked up first, very little sample is lost in the remaining 2 μL.

29. With an expandable pipettor, it is possible to use the wide channel spacing (9.0 mm) to remove the samples from a row of the 96-well plate, then collapse the channel spacing to an appropriate distance (e.g., 4.5 mm for E-gels) for dispensing the samples into the gel wells.
30. Gels enclosed in a plastic cassette are especially easy to load with a multichannel pipettor because the top plastic plate protects the surface of the gel from accidental puncture while aligning the tips with the wells. Some caution must be exercised to avoid puncturing the agarose at the bottom of the gel wells as this can lead to sample leakage between the gel and the lower plastic plate. Gently pressing the tips against one side of the wells will help stabilize the pipettor while dispensing the samples and assure that little residual sample will remain attached to the tips.
31. With more sizes and regular intervals, the ladder becomes more accurate. However, the creation of these ladders is very time consuming because many alleles must be amplified for each locus, the products must be quantified and the components assembled in appropriate ratios.
32. Data from small peaks (< 100 RFU) should be repeated with more sample (a lower dilution of the PCR or, if the amplification was not sufficiently productive, with a new amplification reaction with more template).

INDEX